Recent Results in Cancer Research

Fortschritte der Krebsforschung

Progrès dans les recherches sur le cancer

VI

Edited by

V. G. Allfrey, New York · M. Allgöwer, Chur · K. H. Bauer, Heidelberg · I. Berenblum, Rehovoth · F. Bergel, London · J. Bernard, Paris · W. Bernhard, Villejuif
N. N. Blokhin, Moskva · H. E. Bock, Tübingen · P. Bucalossi, Milano · A. V. Chaklin, Moskva · M. Chorazy, Gliwice · G. J. Cunningham, London · W. Dameshek, Boston
M. Dargent, Lyon · G. Della Porta, Milano · P. Denoix, Villejuif · R. Dulbecco, San Diego · H. Eagle, New York · R. Eker, Oslo · P. Grabar, Paris · H. Hamperl, Bonn
R. J. C. Harris, London · E. Hecker, Heidelberg · R. Herbeuval, Nancy · J. Higginson, Lyon · W. C. Hueper, Bethesda · H. Isliker, Lausanne · D. A. Karnofsky, New York · J. Kieler, København · G. Klein, Stockholm · H. Koprowski, Philadelphia · L. G. Koss, New York · G. Martz, Zürich · G. Mathé, Paris · O. Mühlbock, Amsterdam · G. T. Pack, New York · V. R. Potter, Madison · A. B. Sabin, Cincinnati · L. Sachs, Rehovoth · E. A. Saxén, Helsinki · W. Szybalski, Madison
H. Tagnon, Bruxelles · R. M. Taylor, Toronto · A. Tissières, Genève · E. Uehlinger, Zürich · R. W. Wissler, Chicago · T. Yoshida, Tokyo · L. A. Zilber, Moskva

Editor in chief

P. Rentchnick, Genève

Springer Science+Business Media, LLC 1966

Malignant Transformation
by Viruses

Edited by

W. H. Kirsten

With 39 Figures

Springer Science+Business Media, LLC 1966

Symposium on Malignant Transformation by Viruses
The University of Chicago School of Medicine
February 26th and 27th, 1966

Sponsored by the Swiss League against Cancer

ISBN 978-3-642-87404-8   ISBN 978-3-642-87402-4 (eBook)

DOI 10.1007/978-3-642-87402-4

© 1966 Springer Science+Business Media New York
Originally published by Springer-Verlag New York Inc. in 1966
Softcover reprint of the hardcover 1st edition 1966

Library of Congress Catalog Card Number: 66-23138

Title No. 7470

Preface

The position of "Cancer Teaching Coordinator" at The University of Chicago has been a consistently rewarding one because of the enthusiasm and support of the faculty and the students. This volume is the result of the second of two recent intensive teaching sessions which have been planned and implemented by the group which forms the Cancer Coordinator's Advisory Committee. The first of these teaching sessions was held in early March of 1964 and was entitled "LEUKEMIA, A Current and Forward Look." It attracted overflow attendance from the students and staff of this medical institution augmented by members of the other medical centers in Chicago. It was a stimulating and instructive colloquium, and the only regret we heard expressed afterward was that we had not arranged for publication of the many excellent presentations.

One of the events commemorating the seventy-fifth anniversary of the founding of The University of Chicago was this symposium on malignant transformations. This time the Committee advised us to plan on speedy publication, and, logically, it chose Dr. Werner Kirsten, a member of our faculty and an active and effective investigator in this general field of endeavor, to serve as editor of the volume.

Again, two of the same ingredients for effective instruction were blended: an excellent group of scientists presenting their latest work and a fine, attentive audience of students and staff. This time we were also privileged to have the valuable assistance of Miss Judith Strup, who labored diligently in making the necessary arrangements for several weeks prior to the conference and later in transcribing the round-table and the summary. We also want to acknowledge the splendid help and cooperation of Springer-Verlag New York Inc., as well as the effective work of Mrs. Nathalie Weil, Mrs. Brenda Pickens and Miss Suzanne Larsen, all of whom used their skills to get the manuscript into submittable form in record time after the conference.

In this and in our many other ventures, the Cancer Coordinator recognizes the contributions of his Advisory Committee and thanks them for their help. They are as follows:

1. Clifford W. Gurney, Professor of Medicine and Physiology, Markle Scholar, and Head of the Section of Hematology in the A.E.C. Argonne Cancer Research Hospital.

2. Melvin L. Griem, Associate Professor of Radiology in the A.E.C. Argonne Cancer Research Hospital.

3. Paul V. Harper, Professor of Surgery and Associate Director of the A.E.C. Argonne Cancer Research Hospital.

4. Elwood V. Jensen, American Cancer Society—Charles Hayden Foundation Research Professor of Physiology in the Ben May Laboratory for Cancer Research.

5. Werner H. Kirsten, Associate Professor of Pathology and Pediatrics and N.C.I. Career Development Awardee.

6. John V. Prohaska, Professor of Surgery.

7. Henry Rappaport, Professor of Pathology and Director of Surgical Pathology.

ROBERT W. WISSLER, PH.D., M.D.
Cancer Coordinator and
Professor of Pathology,
University of Chicago
School of Medicine and
in the A.E.C. Argonne
Cancer Research Hospital

April, 1966

Contents

The Characteristics of Malignant Transformation by Viruses

Chairman: R. Haselkorn

Genetics and Immunology of Malignant Transformation by Viruses

Chairman: B. Roizman

SIGNIFICANCE OF MALIGNANT TRANSFORMATION IN RELATION
TO HUMAN NEOPLASMS

Chairman: J. L. Melnick

List of Participants

BLACK, PAUL H., M.D.
Laboratory of Infectious Diseases
National Institute of Allergy and Infectious Diseases
National Institutes of Health
Bethesda, Maryland

DAWE, CLYDE J., M.D.
Laboratory of Pathology
National Cancer Institute
National Institutes of Health
Bethesda, Maryland

DEFENDI, VITTORIO, M.D.
The Wistar Institute
Philadelphia, Pennsylvania

DE OME, KENNETH B., PH.D.
Cancer Research Genetics Lab
Department of Zoology
University of California
Berkeley, California

GIRARDI, ANTHONY J., PH.D.
The Wistar Institute
Philadelphia, Pennsylvania

GREEN, HOWARD, M.D.
Department of Pathology
New York University School of Medicine
New York, N.Y.

HABEL, KARL, M.D.
Laboratory of Biology of Viruses
National Institute of Allergy and Infectious Diseases
National Institutes of Health
Bethesda, Maryland

HASELKORN, ROBERT, PH.D.
Department of Biophysics
University of Chicago
Chicago, Illinois

HILLEMAN, MAURICE R., PH.D.
Merck Institute for Therapeutic Research
West Point, Pennsylvania

HUEBNER, ROBERT J., M.D.
Laboratory of Infectious Diseases
National Institute of Allergy and Infectious Diseases
National Institutes of Health
Bethesda, Maryland

KIRSTEN, WERNER H., M.D.
Department of Pathology
University of Chicago
Chicago, Illinois

LASFARGUES, ETIENNE Y., D.V.M.
Department of Microbiology
Columbia University College of Physicians & Surgeons
New York, N.Y.

MACPHERSON, IAN A., B.Sc., PH.D.
Experimental Virus Research Unit
Institute of Virology
University of Glasgow
Glasgow, Scotland

MELNICK, JOSEPH L., PH.D.
Department of Virology and Epidemiology
Baylor University School of Medicine
Houston, Texas

MOORHEAD, PAUL S., PH.D.
The Wistar Institute
Philadelphia, Pennsylvania

NICHOLS, WARREN W., M.D., PH.D.
Department of Cytogenetics
South Jersey Medical Research Foundation
Camden, New Jersey

RABSON, ALAN S., M.D.
  Pathologic Anatomy Department
  National Cancer Institute
  National Institutes of Health
  Bethesda, Maryland

RAPP, FRED, PH.D.
  Department of Virology and Epidemi-
    ology
  Baylor University School of Medicine
  Houston, Teaxs

ROIZMAN, BERNARD, PH.D.
  Department of Microbiology
  University of Chicago
  Chicago, Illinois

SABIN, ALBERT B., M.D.
  Virus and Cancer Research Laboratory
  Department of Pediatrics
  University of Cincinnati College of
    Medicine
  Cincinnati, Ohio

TEMIN, HOWARD M., PH.D.
  McArdle Laboratory for Cancer Re-
    search
  University of Wisconsin
  Madison, Wisconsin

YERGANIAN, GEORGE, PH.D.
  Children's Cancer Research Hospital
  Harvard University Medical School
  Boston, Massachusetts

Introduction

Cancer research has proceeded at a rapid pace over the last decade. One particularly significant accomplishment has been the increasing knowledge of the initial step in the formation of cancer cells from normal progenitor cells. Perhaps no model system has yielded more basic understanding than studies with tumor viruses in tissue culture. Although such attempts date back to Carrel's work with Rous sarcoma virus and blood monocytes in 1925, the results were not generally accepted, nor was the experimental model widely used. It was not until some twenty-five years later that investigations into malignant transformation in tissue culture were re-opened by the use of "old" and "new" tumor viruses. The significance of virus-induced neoplastic changes in cell cultures can hardly be overestimated.

The Cancer Coordinator at The University of Chicago with the help of his Advisory Committee made plans to hold a teaching symposium on the subject of viral malignant transformation as a part of the 75th anniversary of the University of Chicago. The objectives of the conference were to survey the recent accomplishments in this field and to present for the medical students and the house staff the different morphologic, cytochemical, cytogenetic and immunologic aspects of this topic. Leading scientists in these fields were invited to present general reviews and their latest results. Each speaker was followed by an invited discussant.

Because of the increasing importance of antigenic differences between normal and virus-transformed cells, Dr. Maurice Hilleman was asked to analyze the current feasibility of an "Immunologic Attack on Neoplasia." A round-table discussion was arranged so that the main speakers could exchange additional information and could answer questions from the audience. This was under the especially stimulating leadership of Dr. Robert Huebner. We were particularly fortunate that Dr. Albert Sabin agreed to summarize the results of the symposium.

The symposium was held on February 26 and 27, 1966, at the School of Medicine, The University of Chicago. Many interested scientists attended the conference. The response from our graduate and postgraduate students for whom this conference was arranged was most gratifying. The symposium was supported in part by the Rosa Kuhn Levy Fund and by the Undergraduate Cancer Teaching Grant of the National Cancer Institute #T2 CA 480–18.

We are grateful to all who have made this conference possible.

WERNER H. KIRSTEN

Malignant Transformation and Reversion in Virus Infected Cells

By

IAN MACPHERSON

Institute of Virology, Glasgow, Scotland

The transformation of cells by some oncogenic viruses has been shown to be both direct and rapid, and a study of this interaction *in vitro* probably offers the best opportunity for elucidating the early events of neoplastic conversion.

A classification of oncogenic viruses into those containing DNA or RNA is, to a large extent, mirrored by the type of interaction they make with the cells they transform. The persistence of some virus specific influence in transformed cells is the common finding in the systems studied so far. In cells rendered neoplastic by RNA-containing viruses of the murine leukaemic and avian leukosis groups, virus continues to be synthesized in the cytoplasm and released by budding from the cytoplasmic membrane. Large amounts of virus are present in extracts of tumours and also in transformed cells grown *in vitro*. Specific tumour antigens have also been shown to be produced in some cases.

In cells transformed by the small DNA viruses such as polyoma, SV40 and the oncogenic adenoviruses, infectious virus is no longer detectable in cell extracts and can only be induced in exceptional cases. However, virus-specific antigens may be detected in these cells by complement-fixation, immunoflorescence and transplantation immunity tests.

A third category must be provided for by the Bryan high titre stain (BH) of Rous sarcoma virus (RSV) in its interaction with chicken cells and also for some other strains of RSV such as the Schmidt-Ruppin strain (SR) in mammalian cells. The type of interaction resulting from transformations in these systems resembles that produced by small DNA viruses in that infectious virus either is not formed by the transformed cells or is produced in very small amounts. They do, however, differ from the DNA viruses in that infectious virus may be obtained from these cells. In the case of chicken cells transformed by BH-RSV, the cells are super-infected with a helper virus, i.e., another virus of the avian leukosis group that is capable of providing the "defective" BH-RSV with a protein coat. In the case of mammalian cells transformed by SR-RSV, infectious virus is revived by culturing the cells along with chick-embryo cells or by implanting large numbers of the cells into chickens. When this is done the mammalian tumour cells make contact with the chicken cells and the RSV genome is, in some way, transferred to them. Since SR-RSV is non-defective in chicken cells, complete infectious virus is formed. This leads to the formation of transformed foci *in vitro* and to virus-releasing tumours *in vivo*.

1

IAN MACPHERSON

TABLE I

TRANSPLANTABILITY OF BHK21/13 CELLS AND THEIR POLYOMA AND SR-RSV
TRANSFORMED DERIVATIVES

Cells	Number of cells required to produce tumours in 50% of hamsters 10 weeks after cheek-pouch inoculation					
	10^1	10^2	10^3	10^4	10^5	10^6
BHK21 – clone 13						
" Polyoma transformed						
" SR-RSV transformed 1						
" SR-RSV transformed 2						
" SR-RSV transformed 3						
" SR-RSV "revertant" 1 (derived from SR-RSV-2)						
" SR-RSV "revertant" 2 (derived from SR-RSV-2)						

The persistence of virus or virus-specific effects in transformed cells suggests that the continued presence of viral genes may be necessary for the cell to express its neoplastic properties. The experiments described here provide evidence to support this idea.

The results were the early outcome of a comparative study of transformations induced by a DNA virus (polyoma) and an RNA virus (RSV) in a cloned line of hamster fibroblasts (BHK21/13). It was hoped that in such a study the recognition of points of similarity and difference would lead to a better understanding of the mechanisms by which these different viruses induce cell transformation.

At the time this study was commenced we had analysed the interaction of polyoma virus in BHK21/13 cells in some detail (7, 15, 16). We had developed two basic techniques for the recognition of transformed cells. The first technique involved the plating of infected cells at suitable dilutions in medium to ensure that the individual infected cells gave rise to discrete colonies (7). Transformed colonies were easily recognized because the cells they contained grew in disarray and piled up on each other. The colonies derived from untransformed cells were composed of cells growing in parallel array. The second technique developed in our studies with polyoma resulted from the observation that transformed cells readily grew in fluid suspension culture, whereas the untransformed cells did not (8, 10). Making use of this fact, we devised a selective assay based on the agar suspension culture method developed by Sanders and Burford (11). Only transformed cells have a high colony-forming capacity in this type of culture.

The Schmidt-Ruppin (SR) strain of RSV was selected as a suitable RNA virus for our comparative study because it has been shown to be active in causing tumours in mammals, especially hamsters, and also capable of transforming mammalian cells *in vitro* (1, 5). When BHK21 cells were infected with SR-RSV and cultured in agar suspension, a few colonies grew out. These had a diffusing morphology unlike the compact, smooth-edged colonies induced by polyoma virus. The colonies were picked from the agar and grown on glass. All the isolates grew in disarray like polyoma-transformed cells (Figs. 1 and 2). When these cultures were plated to produce discrete colonies on glass, the majority of the colonies were of the transformed type, but in the progeny from some agar isolates, a small per cent of the colonies was composed of cells growing in parallel array (Fig. 2). It was possible that these colonies with parallel orientation were derived from untransformed cells taken from agar with the RSV-transformed colony. To eliminate this possibility and to establish that the colonies with parallel orientation had at one time a transformed ancestor, two cycles of single-cell cloning by micro-manipulation were carried out on cells from a transformed colony. When the progeny of these clones were plated on glass, from 1 to 10% of the colonies formed had well-marked parallel orientation. Other colonies with intermediate morphologies were always present. There can be little doubt that the colonies with parallel orientation had a transformed cell as an ancestor (9).

Single cell clones were then prepared from one "revertant" colony and their characteristics examined in parallel with the cloned populations of their transformed precursors.

Transplantation of the cells subcutaneously in the backs or intradermally in the cheek pouches of young adult hamsters showed that the transformed cells were as transplantable as some lines of polyoma transformed cells and that the revertant cells had lost this enhanced transplantability.

Transplantations of the cells into day-old chickens or young adult birds from commercial sources and also from a flock of inbred brown Leghorn chickens known to be free from avian visceral lymphomatosis, all produced similar results. If 10^6 to 10^7 transformed cells were injected intramuscularly, about half the birds developed firm sarcomas at the site of inoculation within 3 weeks. Cells irradiated with 1,500 r were at least as successful at initiating tumours. With one exception, the revertant cells did not induce tumours in chickens. The exception occurred in one batch of 30 commercial chickens inoculated when one day old with 6×10^6 cells. One bird developed a tumour after 28 days. Five tumours induced by the transformed cells have been cultured and the karyotypes of the cells growing out examined. Only cells with chicken chromosomes were detected. The sera from four birds of the avian lymphomatosis-free flock, bearing large tumours induced by transformed cells, were tested for their neutralising activity against SR-RSV. Pock formation by the SR-RSV strain on the chick embryo chorioallantois was strongly inhibited by all four sera.

Extracts of the transformed cells prepared by freezing and thawing or by ultrasonication have failed to induce tumours in chickens. Individual chickens have been inoculated with extracts of as many as 4×10^7 cells and not developed tumours.

The complement-fixing activity of transformed and reverted cell extracts against serum from hamsters bearing SR-RSV tumours was also examined. The hamster tumours had been induced initially by infecting new-born hamsters with SR-RSV. These tumours were then passed in adult hamsters. When large tumours had developed, blood was taken. As Huebner et al. (4), have shown, this serum fixes complement when reacted with an antigen found in crude extracts of cells infected with members of the avian leukosis virus complex. This has been called the COFAL test. Transformed and reverted cells were frozen and thawed or ultrasonicated as 10% suspensions and reacted with COFAL positive hamster serum. All the transformed clones were found to possess COFAL antigen in significant amounts but none could be detected in the revertant cells.

Transformed cells grown on glass and then returned to agar suspension culture had plating efficiencies of about 5%. The revertant clones did not plate in agar any better than low passage BHK21 cells, i.e., less than 0.01 per cent of the plated cells formed progressively growing colonies when 10^4 cells were plated in each culture.

Two revertant clones were infected with the same stock of SR-RSV that caused their original transformation and then plated in agar suspension culture. No transformation occurred in this experiment nor in a repeat experiment. It is not clear why the revertant cells are resistant to transformation.

With the exception of reversion, the behaviour of the SR-RSV cells with respect to the status of the virus genome they contain, corresponds to that described by Svoboda and his colleagues for the rat cells (XC) transformed by the Prague strain of RSV (13, 17).

In our system it is clear that morphological reversion is correlated with reduction in transplantability and the loss of detectable viral genome as shown by the negative COFAL test and the inability of the majority of revertant cells to induce tumours in chickens. This loss may only be a quantitative effect and "reversion" may be apparent when the virus-coded products in a cell have fallen below a certain level. Revertants have bred true to type and no transformed colonies have been found in platings of revertant cells carried in serial culture. The rare appearance of a tumour in chickens inoculated with revertant cells may be due to the occasional resurgence of RSV genome in some cells.

Returning to the association of polyoma virus with the cell it transforms, we find that all the available evidence points to an intimate integration of at least part of the virus genome with that of the cell. The best evidence for this comes from the experiments of Sjögren (14) who showed that polyoma-induced mouse tumour cells do not lose their virus-specific tumour antigen when they are passaged for as many as 42 times in mice immunised against this antigen. In this experiment the tumour cells for each passage were obtained from a mouse of the previous passage bearing a tumour induced by the smallest dose of cells. The absence of loss mutants under such strong selective pressure suggests that the continued existence of the tumour antigen or the associated virus information is necessary for the viability or transplantability of the cell. Experiments in our laboratory involving the examination of many thousands of platings of polyoma-transformed BHK21/13 cells, growing in a variety of media, have never resulted in the appearance of any colonies with untransformed morphology. The small DNA viruses replicate in the nucleus during their vegetative cycle, and the idea that in the cells they transform, their genome has made an intimate integration with the host cell DNA, remains the most favored hypothesis to explain the persistence of virus effects in these cells.

It is not difficult to conceive of a model in which replicating RNA particles can be lost from a cell by the simple dilution of the cytoplasm in cell division, especially if cessation or retardation of the RNA's replication occurs.

A useful analogy comes from the work of Beale (2) and Gibson and Sonneborn (3). They have found that the persistence of bacteria-like entities called "mu particles" in the cytoplasm of paramecia is dependent on a stable, gene-initiated RNA. This RNA particle, which may be capable of slow replication in paramecia is called the metagon. When the host cell determinants are lost by mating, the metagons carried into the non-permissive host cell persist for a limited period. Two factors influence the duration of this period. One is the replication of metagon particles and the other is their depletion by cytoplasmic dilution occurring as a result of cell division. The latter effect is greater, and metagons are lost from cells after about 15 divisions. Mu particles disappear from the cells shortly after they have been deprived of their supporting metagons. Another species of protozoon, Didinium, feeds on paramecia. If it ingests paramecia with metagons, it can be shown that the metagons will replicate indefinitely in their new host. Thus, in Didinia, the metagon acts like an RNA virus, but when present in the cytoplasm of the non-permissive paramecia (although it still functions in its capacity as an mu particle maintainer), it is gradually lost. RSV can be likened to a replicating and stable messenger RNA. When it infects chicken cells and some mammalian cells such as the RSV-transformed rat line, XC, in which the RSV-determined characters are stable, it is behaving like the metagon in Didinia. When it infects BHK21/13 cells and is lost on subculture, it behaves like the metagon in a non-permissive paramecium. Perhaps the appropriate genetic support for RSV-RNA replication may be absent in BHK21/13 cells or is rarely found in these cells and rapidly lost by back mutation. Alternatively, the combination of the poorly replicating Schmidt-Ruppin genome with the rapidly multiplying BHK21/13 cells may tend to result in the loss of the virus genome from the cells. Thus, the appearance of revertants in this system may be due to nothing more mysterious than the fact that these cells have such a high rate of multiplication and in the untransformed or reverted condition they are not at a great growth disadvantage compared with the transformed cells. In other systems of primary cultures of cells, viral transformation usually endows the cell with a greatly enhanced ability to grow in tissue culture by conferring a shorter generation time and a higher plating efficiency. A revertant cell would be at a considerable disadvantage if it regained its original characteristics with reversion and had to compete in mixed culture with transformed cells. In such a situation it may not be detectable.

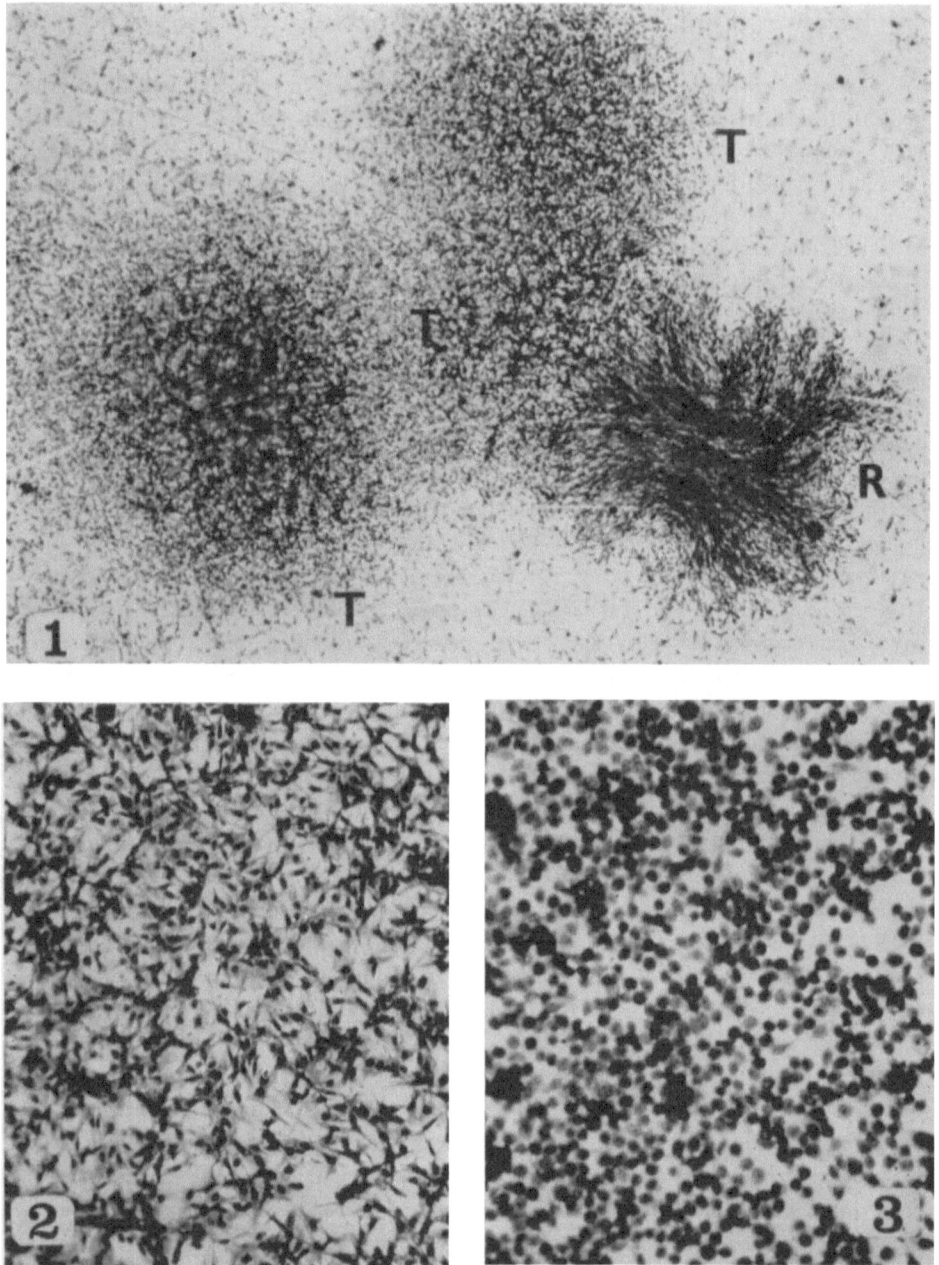

Figs. 1, 2 and 3.

In order to investigate further the possibility that reversion may be a peculiarity of SR-RSV virus, especially in combination with the BHK21/13 cells, attempts were made to produce transformed BHK21/13 with Bryan high titre virus.

Bryan RSV with the RAV 1 pseudo type was obtained from two sources. One stock failed repeatedly to induce agar colony formation in BHK21/13 cells. The other stock, which contained approximately the same titre of focus-forming activity in chick embryo cells as the inactive stock, produced a low rate of transformation. When the agar colonies were isolated and grown on glass, the cells were found to be epithelioid or rounded (Fig. 3). Single cell clones were prepared from these isolates and these were first examined by the COFAL test to determine whether the cells had in fact been infected with RSV. Of 8 clones tested, all were positive. When these cells were plated to produce discrete colonies, none of the clones was found to contain revertants. These platings have been repeated several times with the same results. It is possible that the stability of this transformation is due to the Bryan RSV strain's ability to replicate its genome more successfully than the SR strain; e.g., when these strains are grown in chick embryo cells with RAV 1 helpers, the Bryan strain always produces about 100-fold more focus-forming units of virus than the SR strain from the same number of cells.

Hanafusa (personal communication) has recently shown that the ability of RSV to induce mammalian tumours and presumably to transform mammalian cells, is chiefly determined by the viral envelope. Thus, the enclosure of the genome of SR-RSV within a coat of RAV 1 specificity rendered it incapable of inducing tumours in new-born hamsters. On the other hand, the inclusion of the Bryan strain genome within an envelope supplied by a newly isolated helper virus (RAV-50) makes it an efficient oncogenic agent when injected into new-born hamsters, e.g., when $0.6 - 5.1 \times 10^3$ focus-forming units of Bryan RSV (RAV-50) were injected into day-old hamsters, 21 of 24 developed tumours within 15 days.

Thus, it seems that when the Bryan genome is introduced into the hamster cell, it is just as capable as the SR genome of causing a neoplastic transformation and may be more efficient in maintaining it. It is of interest to note that the transformations produced by each of the two virus strains in chicken and hamster cells are similar. The SR strain induces large diffusing foci in chick embryo cells and converts the BHK21/13 cells into randomly growing fibroblasts. The Bryan strain, both in chick embryo cells and in the BHK21/13 cells, makes the cells round or epithelioid.

The example given here, of reversion in a neoplastic cell, is certainly not without precedents. Seiler-Aspang and Kratochwil (12) have shown that epithelial tumour cells induced by benzpyrene in the newt may spontaneously differentiate into normal tissues. Kleinsmith and Pierce (6) have induced tumours in mice with single embryonal carcinoma cells and demonstrated their ability to differentiate into benign tissues.

The observation of reversion in the SR-RSV transformed BHK21/13 cells raises more questions than it answers, and these are the subject of current investigations in our laboratory. Attempts are being made to find the same phenomenon in other cell systems, in particular lines of inbred rat cells and in cells transformed *in vivo*.

References

1. AHLSTRÖM, C. G., and FORSBY, N.: Sarcomas in hamsters after injection with Rous chicken tumor material. J. Exp. Med. 115, 839–862 (1962).
2. BEALE, G.: Genes and cytoplasmic particles in Paramecium. *In:* Symposium on cellular control mechanisms in cancer. P. Emmelot, O. Mühlbock (eds.). Amsterdam: Elsevier Publ. Co., 1964, pp. 8–18.
3. GIBSON, I., and SONNEBORN, T. M.: Is the metagon on n-RNA in Parmecium and a virus in Didinium? Proc. Nat. Acad. Sc. 52, 869–876 (1964).
4. HUEBNER, R. J., ARMSTRONG, D., OKUYAN, M., SARMA, P. S. and TURNER, H. C.: Specific complement-fixing viral antigens in hamster and guinea pig tumors induced by the Schmidt-Ruppin strain of avian sarcoma. Proc. Nat. Acad. Sc. 51, 742–750 (1964).
5. JENSEN, F. C., GIRARDI, A. J., GILDEN, R. V., and KOPROWSKI, H.: Infection of human and simian tissue cultures with Rous sarcoma virus. Proc. Nat. Acad. Sc. 52, 53–59 (1964).
6. KLEINSMITH, L. J., and PIERCE, G. B.: Multipotentiality of single embryonal carcinoma cells. Cancer Res. 24, 1544–1552 (1964).
7. MACPHERSON, I.: Reversion in hamster cells transformed by Rous sarcoma virus. Science 148, 1731–1733 (1965).
8. MACPHERSON, I., and MONTAGNIER, L.: Agar suspension culture for the selective assay of cells transformed by polyoma virus. Virology 23, 291–294 (1964).
9. MACPHERSON, I., and STOKER, M.: Polyoma transformation of hamster cell clones—an investigation of genetic factors affecting cell competence. Virology 16, 147–151 (1962).
10. MONTAGNIER, L., and MACPHERSON, I.: Croissance sélective en gélose de cellules de hamster transformées par le virus du polyome. Compt. rend. Acad. Sc. 258, 4171–4173 (1964).
11. SANDERS, F. K., and BURFORD, B. O.: Ascites tumours from BHK.21 cells transformed *in vitro* by polyoma virus. Nature 201, 786–789 (1964).
12. SEILER-ASPANG, F., and KRATOCHWIL, K.: Induction and differentiation of an epithelial tumour in the newt (Triturus cristatus). J. Embryol. Exp. Morph. 10, 337–356 (1962).
13. SIMKOVIC, D.: Interaction between mammalian tumor cells induced by Rous virus and chicken cells. Nat. Can. Inst. Monograph 17, 351–364 (1964).
14. SJÖGREN, H. O.: Studies on specific transplantation resistance to polyoma-virus-induced tumors. IV. Stability of the polyoma cell antigen. J. Nat. Can. Inst. 32, 661–666 (1964).
15. STOKER, M., and ABEL, P.: Conditions affecting transformation by polyoma virus. Cold Spring Harbor Symposium Quantitative Biology. 27, 375–386 (1962).
16. STOKER, M.: Proceedings 6th Canadian Cancer Conference (*in press*).
17. SVOBODA, J.: Malignant interaction of Rous virus with mammalian cells *in vivo* and *in vitro.* Nat. Can. Inst. Monograph 17, 277–298 (1964).

Discussion

By Howard Green

New York University School of Medicine
New York, New York

The discovery that oncogenic viruses produce transformations *in vitro* was first made using cell populations recently derived from disaggregated tissues. Such cell populations have the properties characteristic of cell strains as defined by Hayflick; they have relatively short culture life-times, rather low plating efficiency, and are not completely homogenous. They can be cloned only with difficulty and quantitative studies of transformation frequency are not easy to perform.

An important advance was therefore made by Macpherson and Stoker when they developed the BHK 21 line and introduced it for the study of polyoma transformation. This line arose spontaneously (a very unusual event in the Syrian hamster) from disaggregated kidney tissue and has properties characteristic of established lines, such as infinite growth potential and high plating efficiency. However, it preserves a highly oriented pattern of growth in culture which is probably due to the fact that this cell type is highly susceptible to contact inhibition of cell movement. The loss of this character in colonies developing after viral infection is used in scoring the transformation.

In addition to this change, virus-transformed BHK 21 acquire the ability, not possessed to any degree by the parent line, to grow from single cells suspended in agar. Finally, they acquire increased transplantability *in vivo*. These changes developed together in the transformants described by Dr. Macpherson. The change of colonial morphology and the ability to grow in agar both result from a release from cellular inhibition mechanisms and are therefore manifestations of the same process that gives rise to neoplastic properties.

Judging by these criteria, the BHK 21 line is transformed by both polyoma virus and the Schmidt-Ruppin strain of Rous sarcoma virus. Though the transformants are quite similar in appearance and properties, one type never gives rise to revertants, while the other does so with a high probability. This suggests that there may be an important difference in the mechanism of transformation by the two viruses. The fact that the Bryan strain transformants do not give rise to such revertants does not argue against the idea that the revertants reflect a difference between the behavior of the DNA and RNA viruses. Since the revertants from Schmidt-Ruppin transformants seem to be due to loss of the Schmidt-Ruppin Rous genome, an event which would require that the viral genome replicate at a rate somewhat slower than that of the cell which is to revert, it seems likely that the Bryan strain simply replicates too rapidly for its transformants ever to become virus-free.

We may ask in what other respects DNA viruses and RNA viruses might differ with regard to transformation. I shall refer to an example of a DNA virus transformation, namely, to that occurring in the 3T3-SV40 system, extensively studied by Dr. George Todaro and myself. The system is quite analogous to the polyoma-BHK system described by Dr. Macpherson.

First, as Dr. Macpherson pointed out, the Rous virus may always be obtained from transformed cells, while with a few possible exceptions, transformants produced by DNA viruses can be made virus-free. In the case of SV40-3T3 interaction, there is probably no viral replication at all, and virus-free clones of transformants may be isolated quite soon after infection. In other systems where there is viral replication, such as the polyoma-mouse cell interaction, it is more difficult to obtain virus-free clones or transformants. However, in all of these cases, the virus-free lines do retain the transformed phenotype.

The 3T3 cell can remain viable for a long time in the non-growing state. It has been possible to show that non-growing cells are not transformable by SV40 virus. Although direct evidence is still lacking, it seems most likely that the cell must be synthesizing DNA in order to be affected by the virus, either because integration of viral DNA is involved, or because only replicating cellular DNA is vulnerable to virus-induced damage. There has been some experimental evidence in the past linking Rous sarcoma virus replication to the host genome, but I wonder if this evidence can be regarded as conclusive. Evidence that DNA synthesis is necessary for Rous virus replication had been found by Dr. Temin and by Dr. Bader, but it is not known that host cell DNA synthesis is involved. Possibly, the Rous sarcoma virus should be regarded as episome-like. If Rous sarcoma virus has no need to interact with cellular DNA, non-growing cells might be fully transformable.

In the SV40-3T3 system, one cell generation in the presence of the virus is sufficient to fix the transformed state. Though transformation is quite effectively prevented by interferon when added immediately after infection, by one cell generation after the infection, the addition of interferon is ineffective. This is consistent with an irreversible viral effect on the host cell genome occurring during the first cell division. In the case of the Rous transformants the situation may be different: possibly interferon could induce revertants in Dr. Macpherson's Schmidt-Ruppin transformants at a rate much higher than the spontaneous reversion rate, and might even do so in the Bryan transformants which do not revert spontaneously. Other antiviral agents such as those specific for the viral RNA polymerase might also have such an effect. Dr. Macpherson has referred to the rapid growth of the BHK 21 cell which is important in diluting out the viral genome. The rapid growth rate may also be important in reducing the rate of replication of the viral genome, for it has recently been shown by Carp and Gilden that the rate of SV40 production in growing cells may be only 1% of the rate of production in non-growing cells. Evidently, growing cells are able to suppress viral replication in certain instances, as viruses are able in others to suppress host cell DNA synthesis.

Again in the case of 3T3, the transformed phenotype requires several cell generations for expression. If this is necessary in order to dilute out some normal cellular constituent no longer made in the transformant, Rous transformants might show a similar behavior. On the other hand, these cell generations might be involved in processes specific for a host cell DNA-viral DNA interaction. In this case, Rous transformants might not show a similar requirement.

I have left out of account one important fact about Rous virus first shown by Temin and Rubin. A cell whose colony forming ability has been destroyed through x-irradiation does not initiate Rous virus replication. On the other hand, once viral multiplication has begun, further viral synthesis is resistant to X-ray. This is rather analogous to the events I have described for the SV40-3T3 interaction, but it is not possible to say yet whether the early stages are really the same in both.

Finally, it is interesting that the revertants described by Dr. Macpherson would not retransform or, presumably, make virus after a fresh infection with the Schmidt-Ruppin strain. Transformants produced by the DNA viruses do not lose the ability to support viral replication on fresh infection, if the parent cells were able to do so. Of course, if the reason for the reversion of the Schmidt-Ruppin transformants is that those cells are variants which replicate the Schmidt-Ruppin strain particularly poorly, then it is to be expected that a fresh infection would be ineffective. However, in that case, an infection of the revertant with the Bryan strain might produce transformation.

Malignant Conversion by Avian Sarcoma Virus [1]

By

HOWARD M. TEMIN [2]

McArdle Laboratory, University of Wisconsin, Madison, Wisconsin

Basic cancer research involves two major questions: how cancer cells arise and how they differ in behavior from normal cells. Study of the interaction of the Rous sarcoma virus and related avian tumor viruses with sensitive chicken and duck cells in cell culture has suggested tentative answers to these questions. It appears that, in this system, cancer cells arise by the addition of genes to the normal cell and that these new genes, when expressed, cause changes in the cell so that the cancer cells require less than normal cells of a protein present in serum which controls cell multiplication. Therefore, under conditions in which this protein is limiting, the cancer cells have a selective advantage over normal cells.

This interaction in cell culture between Rous sarcoma virus and sensitive chicken cells has the following characteristics which distinguish it from most other virus-cell systems:

1. Under proper conditions, most cells exposed to virus are infected.
2. No cells are killed by infection.
3. Most cells infected by virus become, under proper conditions, altered in their properties within a few days after infection. This alteration is called conversion.
4. Virus production and conversion are, under proper conditions, separable—both non-converted, virus-producing cells and converted, non-virus-producing cells exist.
5. Infected cells contain a provirus.
6. The genes in the provirus are always recoverable in infectious virus, even from converted, non-virus-producing cells.
7. Converted cells multiply, under proper conditions, at the same rate as normal cells; under other conditions, at different rates.
8. Converted cells do not form permanent cell lines.

These are characteristics of the interactions of all of the avian sarcoma viruses with sensitive chicken cells. Of course, there are numerous other properties of the viruses such as antigenicity, host range, type of foci produced, presence of con-

---

[1] This investigation was supported in part by United States Public Health Service Research Grant CA-07175 from the National Cancer Institute.

[2] Research Career Development Awardee 10K3-CA-8182 from the National Cancer Institute, National Institutes of Health, United States Public Health Service.

taminating leukosis virus, etc., which are not shared by all of these viruses. The common characteristics, especially no. 3, are not shared by the related avian leukosis viruses or with the possibly similar mouse leukemia viruses. All of these viruses do have in common the following properties: they do not kill cells, they are related to tumorigenesis, their virions have an outer lipid-containing membrane and a large molecule of RNA, and they have, perhaps, an RNA to DNA to RNA type of replication.

The process of infection giving rise to a tumor cell in the Rous sarcoma virus-chicken cell system appears to involve three steps: Formation by the infecting virus of new genes (provirus) in the infected cell, activation of these genes and expression of the gene action. For the tumor cell to produce a tumor also requires an animal with a proper selective environment. A review of the evidence for the first three of these steps has recently appeared (17).

Further evidence relating to the first step, formation of the provirus, has been gained by study of infection of partially synchronized cultures. With such cultures, it has been shown that the length of the latent period decreases as the time of infection approaches the time of mitosis and that blocking of the first mitosis after infection prevents the beginning of virus production. These results suggest that cell division is necessary for formation of the provirus. This hypothesis explains the sensitivity to radiation of the capacity of chicken cells to become infected (8) and the fact that the capacity becomes resistant about 10 hours after infection, after the first cell division would have occurred (19). It also explains why pre-treatment of chicken cells with actinomycin (12, 21), mitomycin (21), BUDR (1) or IUDR (4), prevents infection. This hypothesis could also explain the requirement for new DNA synthesis after infection (2, 13, 15). However, blocking of DNA synthesis in the G2 period, when it does not prevent cell division, still prevents infection. Therefore, infection seems to require cell DNA synthesis in order to allow cell division, viral DNA synthesis (14), and cell division.

Once the provirus has been formed and activated, in some way as yet not understood, many properties of the infected cell may alter. In chicken iris epithelial cells, the following changes occur (3, 16):

1. The shape of the cells changes.
2. The capacity to make pigment is lost.
3. The capacity to make acid mucopolysaccharides is gained.

However, the capacity to make collagen is not gained. These effects show that the effect of the provirus on the host genome is specific: some functions are increased, others decreased, and others remain unaltered.

In order to study these alterations further, a study of the effects of conversion on the production of acid mucopolysaccharides in fibroblasts was carried out (16). It was found that the rate of synthesis of acid mucopolysaccharide, a normal fibroblast function, was increased 2- to 3-fold after infection. This increase took place before the alteration of cell shape occurred; however, neither the concentration of acid mucopolysaccharide in the medium nor the rate of production of acid mucopoly-saccharide appeared to cause the alteration in cell shape.

The increase in acid mucopolysaccharide production did not occur in chicken cells infected with avian leukosis viruses, e.g., in non-converted virus-producing cells.

Hyaluronic acid synthetase is the enzyme which catalyzes synthesis of hyaluronic acid, the acid mucopolysaccharide made by chicken fibroblasts in cell cultures. This enzyme can be isolated from normal and from converted cells (5). It appears to be located, in part, in both normal and converted cells, on the surface of the cell, because whole cells can catalyze the formation of hyaluronic acid from exogenous nucleotide substrates, to which cells are usually impermeable, and because trypsin treatment destroys this part of the enzyme activity. About 36 hours after infection, the activity of this enzyme in infected cells is increased over that in normal cells. This increase is 5- to 10-fold a few days after infection. In order to determine whether the hyaluronic acid synthetase in chicken cells infected with avian sarcoma viruses was different from the enzyme in uninfected chicken cells, the effects of pH and of metals on enzyme activity were measured, as well as the rates of inactivation of the enzyme at different pH's, and the Michaelis constants of the enzyme in infected and uninfected cells. No significant differences in these parameters were found between the enzyme from infected and uninfected cells. These results suggest that the provirus affects the level of a host enzyme rather than causing production of a new one.

Another metabolic alteration in converted cells is an increase in the rate of glycolysis. Under the cell culture conditions used, both converted and normal cells metabolize glucose by glycolysis and not by respiration; however, the converted cells use more glucose and produce more lactic acid than do the normal ones. This increased glycolysis is probably a secondary change rather than a reflection of an increased energy need of converted cells, because normal cells can multiply at about the same rate on galactose as on glucose; however, they use only a small fraction of the galactose as compared with the amount of glucose used.

Although these alterations occur in converted cells, neither they nor the change in shape are an indication that the rate of cell multiplication has changed. As long as the medium is changed frequently, both normal and converted chicken cells continue to multiply and to multiply at the same rate (16). It is only under conditions in which serum is limited that a difference in rate of multiplication is found (18). Under these limiting conditions, brought about by not changing the medium of the cultures, the converted cultures have a saturation density (number of cells and amount of cell protein and DNA per culture when multiplication has ceased) 50% higher than that of normal cultures. The effect of agar or other polyanions is to bind something in serum which is needed for cell multiplication. This same substance is also removed from the medium by cells during growth. It has been hypothesized that normal chicken embryo fibroblasts and converted chicken embryo fibroblasts differ in their requirements for a substance in serum, bound by polyanions, which is needed to carry out cell multiplication (18). Although both normal and converted cells require this substance, the converted cells require less of it per cell division than do normal cells. In actual practice the production by the tumor cells of a toxic substance obscures this difference unless a small amount of agar or other polyanion which binds the toxin is present.

An important test of this hypothesis is afforded by normal and converted duck cells. Here a difference in rate and amount of multiplication occurs even when the medium of the cultures is changed frequently. However, medium in which duck cells have grown is depleted with respect to the serum factor required for multiplication of chicken cells. Furthermore, normal duck cells do not multiply when serum is absent from the medium and do multiply more rapidly when the medium is changed. Therefore, it appears that duck cells also require the same serum factor for cell multiplication as chicken cells, but at a much higher level per cell division. To test this hypothesis the multiplication of mixed cultures of duck and chicken cells was compared with the multiplication of chicken cells alone. It was found that the presence of the duck cells in the cultures inhibited the multiplication of the chicken cells even when the medium was changed frequently.

The multiplication of converted duck cells is also dependent on serum, but, as with converted chicken cells, less is required per cell division than is required by normal cells. Normal human and mouse fibroblasts also seem to be dependent on serum for their multiplication. Since transformation of these and of other cells with oncogenic viruses causes the cells to "pile up" and to have an increased saturation density (6, 9, 10, 11, 20, 22), transformation and conversion may both involve a change in the requirement by cells for this factor in serum.

This factor is found in chicken and calf serums but is not species-specific. It is non-dialyzable and destroyed by incubation with pronase. It is stable to heating at 70° C for 30 minutes, overnight incubation at pH's between 2 and 9.5, but not at pH 10. It can be partially replaced by serum Fraction III (Cohn procedure), but not by Fractions II, III-0, III-2, III-3, IV, IV-1, IV-2, IV-4, IV-7, V, VI or transferrin.

The molecular mechanisms by which the virus brings about all of these changes are not clear. However, there are certain strong similarities to lysogenic conversion in Salmonella (7). In both cases:

1. Properties of the infected cell are altered.
2. The type of alteration is dependent on genes carried by the virus and is changed by mutations in the virus.
3. Stably converted cells carry one or two copies of viral genes in a regularly inherited form (provirus or prophage). These genes are always recoverable in infectious virus.
4. Conversion can occur in the absence of production of infectious virus.
5. Conversion is associated with changes in the level of host cell enzymes.

In both cases of conversion, some genes carried by the virus when introduced into the cell affect one or many of the host cell genes. In viral leukemogenesis it may be assumed that viral genes are present in the infected cell, but because of their location they are activated only in certain types of cells, as in differentiation. In transformation of cells by papova viruses and adeno viruses, it is not yet clear whether the virus acts by adding new genes or by some other mechanism. It is clear, however, that if it does add new genes, fewer are added than is the case in conversion.

References

1. BADER, J. P.: The role of deoxyribonucleic acid in the synthesis of Rous sarcoma virus. Virology 22, 462–468 (1964).
2. BADER, J. P.: The requirement for DNA synthesis in the growth of Rous sarcoma and Rous-associated viruses. Virology 26, 253–261 (1965).
3. EPHRUSSI, B., and TEMIN, H. M.: Infection of chick iris epithelium with the Rous sarcoma virus *in vitro*. Virology 11, 547–552 (1960).
4. FORCE, E. F., and STEWART, R. C.: Effect of 5-iodo-2-deoxyuridine on multiplication of Rous sarcoma virus *in vitro*. Proc. Soc. Exptl. Biol. Med. 116, 803–806 (1964).
5. ISHIMOTO, N., TEMIN, H. M., and STROMINGER, J. L.: Studies on carcinogenesis by avian sarcoma viruses. II. Virus-induced changes in hyaluronic acid synthetase in chicken fibroblasts. J. Biol. Chem. 1966, *in press*.
6. KOPROWSKI, H., PONTEN, J. A., JENSEN, F., RAVDEN, R. G., MOORHEAD, P., and SAKSELA, E.: Transformation of cultures of human tissue infected with simian virus SV40. J. Cell. Comp. Physiol. 59, 281–292 (1962).
7. ROBBINS, R. W., KELLER, J. M., WRIGHT, A., and BERNSTEIN, R. L.: Enzymatic and kinetic studies on the mechanism of o-antigen conversion by bacteriophage E[15]. J. Biol. Chem. 240, 384–390 (1965).
8. RUBIN, H., and TEMIN, H. M.: A radiological study of cell-virus interaction in the Rous sarcoma. Virology 7, 75–91 (1959).
9. SACHS, L., and MEDINA, D.: *In vitro* transformation of normal cells by polyoma virus. Nature 189, 457–458 (1961).
10. SHEIN, H. M., and ENDERS, J. F.: Transformation induced by simian virus-40 in human renal cell cultures. I. Morphology and growth characteristics. Proc. Natl. Acad. Sci. U.S. 48, 1164–1172 (1962).
11. STOKER, M., and MACPHERSON, I.: Studies on transformation of hamster cells by polyoma virus *in vitro*. Virology 14, 359–370 (1961).
12. TEMIN, H. M.: The effects of actinomycin D on growth of Rous sarcoma virus *in vitro*. Virology 20, 577–582 (1963).
13. TEMIN, H. M.: The participation of DNA in Rous sarcoma virus production. Virology 23, 486–494 (1964).
14. TEMIN, H. M.: Homology between RNA from Rous sarcoma virus and DNA from Rous sarcoma virus-infected cells. Proc. Natl. Acad. Sci. U.S. 52, 323–329 (1964).
15. TEMIN, H. M.: The nature of the provirus of Rous sarcoma. Natl. Cancer Inst. Mono. 17, 557–570 (1964).
16. TEMIN, H. M.: On the mechanism of carcinogenesis by avian sarcoma viruses. I. Cell multiplication and differentiation. J. Natl. Cancer Inst. 35, 679–693 (1965).
17. TEMIN, H. M.: Genetic and possible biochemical mechanisms of viral carcinogenesis. Cancer Res. 26, 212–216 (1966).
18. TEMIN, H. M.: Studies on carcinogenesis by avian sarcoma viruses. III. The differential effect of polyanions on the multiplication of normal and converted cells. *Submitted for publication*.
19. TEMIN, H. M., and RUBIN, H.: A kinetic study of infection of chick embryo cells *in vitro* by Rous sarcoma virus. Virology 8, 209–222 (1959).
20. TODARO, G. J., and GREEN, H.: An assay for cellular transformation by SV40. Virology 23, 117–119 (1964).

21. VIGIER, P., and GOLDÉ, A.: Effects of actinomycin D and of mitomycin C on the development of Rous sarcoma virus. Virology 23, 511–519 (1964).
22. VOGT, M., and DULBECCO, R.: Virus-cell interaction with a tumor-producing virus. Proc. Natl. Acad. Sci., U.S. 46, 365–370 (1960).

Discussion

By WARREN W. NICHOLS

South Jersey Medical Research Foundation
Camden, New Jersey

Dr. Temin has presented very important and interesting evidence of somatic mutations in cells infected with viruses that are manifest in both physical changes in shape and biochemical parameters. I would like to present some of our own group's work on the problem of virus-induced somatic mutations.

Mutations can be induced by incorporating new genetic material, altering the genetic material already present, or by a combination of these. Our studies of virus-induced somatic mutation processes have been from the standpoint of observation of chromosomal effects of viruses, and attempts to separate the effects of the added virus genome from the effects of the altered cell genome. Using the Schmidt-Ruppin strain of Rous sarcoma virus in the rat, we have studied chromosome of *in vivo* tumors, tumors carried in tissue culture and diploid rat embryo cells in tissue culture to which the Schmidt-Ruppin virus has been added (5). In all of these systems, we observed chromosome breaks or evidence that chromosome breaks had occurred previously, in that rearrangements were found.

Similarly, when the Schmidt-Ruppin Rous virus was added to human leukocytes, breaks were found (11), but this was not true when the Bryan strain (a strain that usually does not produce tumors in mammals) was added.

Also, studies were performed using measles virus (12, 9, 10, 14) in various *in vivo* and *in vitro* systems where similar and additional observations were made. While other types of changes involving chromosomes could also be produced by various viruses such as chromosome pulverization and mitotic and spindle abnormalities (7), we felt that the most likely to be of mutagenic significance is the simple, or single, break. One would expect that most cells with one of these complete chromosome breaks would not survive and divide again due to loss of genetic material in acentric fragments. However, it is quite conceivable that the incidence of virus-induced breaks is an indicator system for subvisible viable changes similar to X-ray-induced breaks, as an indicator system for mutagenic effects.

The type of breakage seen with virus infection differed markedly from that seen with X-ray. X-ray-induced breaks have a high incidence of chromosome reunion and recombination, while the virus-induced breaks were open breaks with little tendency to reunion. We were aware of work with nucleosides that inhibited the synthesis of DNA and produced breaks morphologically very similar to those seen with virus. This had been demonstrated by Taylor (16) with FUdR in plant material and also by Kihlman (3) with AdR in plant material. Because of this morphologic similarity, we undertook, with Kihlman, a comparative study of some of these nucleosides in mammalian systems in which we had been studying viruses. We found that the breaks produced by araC, AdR and araA (4, 6) could not be distinguished morphologically from the virus-induced breaks. Based on the reasoning that if the breaks observed with viruses were indicators of significant mutations, perhaps these similar breaks induced by inhibitors of DNA synthesis might also be capable of producing somatic mutations of a similar nature to the viruses. Thus, this could offer a method to separate mutations produced by the addition of viral genetic material and those due to altering the genetic material already present in the cell, since virus-like chromosome damage could be produced without addition of virus.

With this in mind, we exposed human diploid cell strains to arabinosylcytosine and observed them for changes similar to those seen with viral transformation (8). Immediately after the inhibitors of DNA synthesis were added, a high incidence of open breaks was found similar to those found in short-term treatments of leukocyte cultures.

After a period of recovery, there were some changes in cell morphology and growth pattern of the cells seen both in the cells growing on the glass and in detailed studies of stained cells (2) as increased cytoplasmic granularity, more conspicuous large nucleoli and enlargement of nuclei.

When the chromosomes were studied in these later passages, changes similar to those of virus transformed cells occurred. These consisted of dicentrics and other rearrangements and recombinations. These observations have many similarities with sequential changes in virus transformed cells, but in addition to changes in cell morphology, growth pattern and sequential chromosome changes, viral transformed cells exhibit continuing mitosis in confluent sheets and an increased growth rate that has not been seen in these cells exposed to inhibition of DNA synthesis.

These cells exposed to nucleosides and analogs also resemble the senescent stage of diploid cultures where chromosome abnormalities and decreased growth rate are seen (17, 15), and the unbalanced growth described in conditions of thymine deficiency in bacterial systems by Cohen and Barner (1), in which DNA synthesis is inhibited and RNA and protein synthesis continue.

Other studies that are being carried out to elucidate the role of virus genome and chromosome damage in the production of somatic mutation include the use of nucleotides and virus in combination (13), where it was found that CTP was synergistic with Schmidt-Ruppin Rous virus in the production of breaks. We are also attempting to inhibit the chromosomal damage induced by the virus by additions of excess of normal deoxyribosides and deoxyribotides, as can be done with the inhibitors of DNA synthesis by adding an excess of the specific deoxyriboside that is affected. It may be possible to prevent the virus-induced chromosome damage and observe the isolated effect of the added virus genome. In addition, radioactively labeled adeno-virus 12 is being used in an attempt to locate the virus genome or its components in the cell. The results are preliminary.

In closing, I should like to ask Dr. Temin a question and make a comment on the serum factor he has described. The question concerns the observed biochemical changes in transformed cells. I wonder if these studies have been performed on mass cultures or if it has as yet been possible to perform them on cloned cells, to determine if each biochemical change occurs in each transformed cell.

The comment on the serum factor that is necessary for the continued growth of cells, relates to an observation we have made. In cultured leukocytes we have been able to increase the number of control and experimentally induced chromosome breaks by refeeding the cells 36 hours after establishing the cultures with media that does not contain serum. It could be that this observed increase in breaks is due to the lack of the same serum factor that Dr. Temin describes.

References

1. COHEN, S. S., and BARNER, H. D.: Studies on unbalanced growth in *E. coli.* Proc. Nat. Acad. Sci. **40**, 885–893 (1954).

2. HENEEN, W. K., and NICHOLS, W. W.: Cell morphology of a human diploid cell strain (WI-38) after treatment with arabinosylcytosine. (*In press*) (1966).

3. KIHLMAN, B. A.: Deoxyadenosine as an inducer of chromosomal aberrations in *Vicia faba.* J. Cell. Comp. Physiol. **62**, 267–272 (1963).

4. KIHLMAN, B. A., NICHOLS, W. W., and LEVAN, A.: The effect of deoxyadenosine and cytosine arabinoside on the chromosomes of human leukocytes *in vitro.* Hereditas **50**, 1, 139–143 (1963).

5. NICHOLS, W. W.: Relationships of viruses, chromosomes and carcinogenesis. Hereditas **50**, 53–80 (1963).

6. NICHOLS, W. W.: *In vitro* chromosome breakage induced by arabinosyladenine in human leukocytes. Cancer Res. **24**, 1502–1505 (1964).

7. NICHOLS, W. W.: The role of viruses in the etiology of chromosomal abnormalities. Amer. Jrl. Hum. Genetics 18, No. 1, 81–92 (1966).

8. NICHOLS, W. W., and HENEEN, W. K.: Chromosomal effects of arabinosylcytosine in a human diploid cell strain. Hereditas **52**, 3, 402–410 (1965).

9. NICHOLS, W. W., LEVAN, A., AULA, P., and NORRBY, E.: Extreme chromosome breakage induced by measles virus in different *in vitro* systems. Hereditas **51**, 380–382 (1964).

10. NICHOLS, W. W., LEVAN, A., AULA, P., and NORRBY, E.: Chromosome damage associated with the measles virus *in vitro.* Hereditas **54**, 101–118 (1965).

11. NICHOLS, W. W., LEVAN, A., CORIELL, L. L., GOLDNER, H., and AHLSTRÖM, C. G.: *In vitro* chromosome abnormalities in human leukocytes associated with Schmidt-Ruppin Rous sarcoma virus. Science **146**, 248 (1964).

12. NICHOLS, W. W., LEVAN, A., HALL, B., and ÖSTERGREN, G.: Measles associated chromosome breakage. Hereditas **48**, 367–370 (1962).

13. NICHOLS, W. W., LEVAN, A., HENEEN, W. K., and PELUSE, M.: Synergism of the Schmidt-Ruppin strain of the Rous sarcoma virus and cytidine triphosphate in the induction of chromosome breaks in human cultured leukocytes. Hereditas (*in press*) (1966).

14. NORRBY, E., LEVAN, A., and NICHOLS, W. W.: The correlation between the chromosome pulverization effect and other biological activities of measles virus preparations. Exp. Cell. Res. (1965) (*in press*).

15. SAKSELA, E., and MOORHEAD, P. S.: Aneuploidy in the degenerative phase of serial cultivation of human cell strains. Proc. Nat. Acad. Sci. **50**, 390–395 (1963).

16. TAYLOR, J. H., HAUT, W. F., and TUNG, J.: Effects of fluorodeoxyuridine on DNA replication, chromosome breakage, and reunion. Proc. Nat. Acad. Sci. **48**, 190–198 (1962).

17. YOSHIDA, M. C., and MAKINO, S.: A chromosome study of nontreated and an irradiated human *in vitro* cell line. Japan. J. Human Genetics **5**, 39–45 (1963).

Epigenetic Factors in the Neoplastic Response to Polyoma Virus

By

Clyde J. Dawe,[1] James H. P. Main,[2] Marilyn S. Slatick and Willie D. Morgan

Laboratory of Pathology, National Cancer Institute, Bethesda, Maryland [3]

One could hardly speak of epigenetic matters before an audience at the University of Chicago without taking cognizance of the outstanding contributions that have come from that institution to the field of epigenetics. Among them, and quite relevant to the work described in this report, are the classical works of Dr. F. R. Lillie and Dr. Hsi Wang on the formation of feathers. In a series of experiments that lead to the delineation of the specific morphogenetic functions of the epithelial and the mesenchymal components of feather follicles (1), Dr. Lillie and Dr. Wang developed a pattern of analysis that has found increasing use in developmental biology, and is beginning to find application in experimental oncology. It is safe to predict that the debt of oncologists to those original investigations will continue to increase.

Review

If oncogenic responsiveness to polyoma virus (PV) were determined entirely by the genetic constitution of the host animal's cells, we should expect tumors to arise from any cell type in those species that develop some tumors in response to this virus. Empirically, this does not hold true. Instead, for each species there is a characteristic group of organs, tissues, and cells from which tumors may be expected to arise in response to PV (2). These specific differences in response pattern might be determined by the combined effects of several factors, an over-simplified list of which includes: (1) varying degrees of exposure of different cells and tissues to the virus; (2) variations in the native and acquired resistance of different cell types to viral infection; (3) local immunologic conditions; (4) factors such as gene activation that condition the host-cell genome and thereby influence the establishment of a cell-virus relationship essential for neoplastic change; and (5) growth, or

[1] The authors are grateful to Mr. Gebhard Gsell for photographic assistance; to Mr. J. P. Summerour for technical assistance in the culture work; and to Mr. J. Albrecht and staff for the histologic preparations.

[2] Present Address: Department of Oral Pathology, School of Dental Surgery, Chambers St., Edinburgh, Scotland.

[3] National Institutes of Health, Public Health Service, United States Department of Health, Education and Welfare.

20

promoting factors that support emergence of a neoplasm by supporting survival and proliferation of virus-altered cells.

Our investigations have been concerned with the last two of these influences. They qualify as epigenetic factors under the broad definition that conceives of epigenetic factors as all those influences that act to permit expression of the latent functional and structural capabilities known to unfold (or develop) during ontogeny. The definition stems back to Aristotle's refutation of the preformation theory of embryogenesis (3). Curiously, one can see a remarkable analogy between the ancient debate on preformation versus epigenesis (4) and the modern concern with sorting out genetic from epigenetic factors in oncogenesis.

More specifically, our investigations have been directed toward analyzing the influence of epithelio-mesenchymal interaction on the neoplastic response of the submandibular gland of the mouse to PV. This gland offers special advantages for such studies because it is susceptible to the oncogenic action of PV, and because its development depends on specific inductive interaction between epithelium and mesenchyme. A brief review of past findings may help to clarify present problems and objectives.

Two technical systems have been developed for these analyses. Though they lack some of the quantitative capabilities desirable for most virologic studies, they have the qualitative capabilities required in developmental biology.

The first system is simply that of organotypic culture (Table I). Starting with submandibular glands of newborn mice, we observed that in the presence of PV, changes occurred *in vitro* that resembled PV-induced tumor development *in vivo*

TABLE I *

RESPONSE OF SUBMANDIBULAR GLAND AT DIFFERENT DEVELOPMENTAL
STAGES TO POLYOMA VIRUS IN ORGANOTYPIC CULTURE

Developmental Stage	Summary of Response
Fully Developed (Adult)	CPE in mesenchyme and epithelium after 7 days. Morphologic transformation and proliferation of epithelium after 14–30 days.
Early Functional (Newborn)	Essentially same as in adult gland.
Embryonic Stages 2 to 5 (13 & 14 day embryos)	Continued development 7–10 days. Then CPE followed by morphologic transformation of epithelium; proliferation after 22–30 days.
Embryonic Stage 1 and Pre-rudimentary Area.	Failure of development or very limited morphogenesis. CPE followed by morphologic transformation of epithelium.

* This table represents a summarization of more detailed data published in reference (7).

(5). Although a proportion of such morphologically transformed cultures gave rise to typical polyoma-type salivary tumors when transplanted to syngeneic newborn mice (5), we could not be absolutely certain that the morphologic transformation was equivalent to neoplastic transformation, because such cultures contained virus, and it was therefore possible that tumor induction in the explants occurred after, rather than before, transfer of the cultures.

The second system consists of infecting tissues *in vitro,* and immediately transferring them subcutaneously to syngeneic newborn mice (6, 7). Oncogenic responsiveness is judged according to the frequency with which tumors appear in the transplants (Table II).

In conjunction with these two systems, a procedure developed by Moscona (8) and applied with modifications to submandibular salivary gland by Grobstein (9, 10), has been used. This consists in separating the epithelium and the capsular mesenchyme from each other by trypsinization. After doing this, one can test for responsiveness of each of the isolated components to PV by studying each in the two systems described above (Table III). These two systems will be referred to as the *in vitro* and the *in vivo* systems respectively.

The following observations were made (7). Morphologic transformation occurred *in vitro* and neoplastic transformation occurred *in vivo* in submandibular salivary gland when PV infection was initiated at any stage of development ranging from Stages 2 and 3 (from 13-day embryos) upward through complete development. At Stage 1 or earlier, no tumors developed in the *in vivo* system, but morphologic transformation was seen *in vitro.* The meaning of the transformation *in vitro* is currently being determined by transferring infected cultures of Stage 1 and prerudimentary areas to new hosts after a 30-day period in culture. Thus far (unpublished data) no

TABLE II *

TUMOR DEVELOPMENT IN POLYOMA VIRUS-INFECTED TRANSPLANTS OF
SUBMANDIBULAR GLANDS IN PROGRESSIVE DEVELOPMENTAL STAGES

Developmental Stage of Transplant	Tumors in Transplants		Tumors in Host's Glands	
	Fraction	Percent	Fraction	Percent
Fully Developed (Adult)	32/41	78	28/41	68
Early Functional (Newborn)	18/30	60	18/30	60
Embryonic Stages 4 and 5 (14-day embryos)	40/56	71	37/56	66
Embryonic Stages 2 and 3 (13-day embryos)	3/28	11	19/28	68
Embryonic Stage 1 and Prerudimentary Area	0/56	0	38/56	68

* This table represents a summarization of more detailed data reported in reference (7).

TABLE III *

FAILURE OF TUMOR DEVELOPMENT IN ISOLATED EPITHELIAL AND MESENCHYMAL
COMPONENTS OF SUBMANDIBULAR RUDIMENTS INFECTED WITH POLYOMA
VIRUS *In Vivo* AND *In Vitro*

Material in Culture or Transplant	Tumors in Transplants		Tumors in Host Glands		Response *In Vitro*
	Fraction	Percent	Fraction	Percent	
Intact Submandibular Rudiments (Stages 2–5)	32/69	46	35/69	51	Morphogenesis. Morphologic transformation and proliferation.
Isolated Submandibular Epithelium	0/26	0	17/26	65	No morphogenesis. No morphologic transformation. No proliferation.
Isolated Submandibular Mesenchyme	0/28	0	17/28	61	Viral CPE. Few atypical cells.
Recombined Epithelium and Mesenchyme	11/47	23	29/47	62	Morphogenesis. Morphologic transformation and proliferation.

* This table represents a summarization of more detailed data reported in reference (7).

neoplasms have appeared in these transplanted transformed cultures, but the observation time (2 to 3 months) is inadequate at this writing.

Relying more heavily on the results *in vivo*, the current interpretation of these findings is that a certain threshold stage of development is necessary before salivary gland epithelium can manifest a neoplastic response to PV. After this threshold stage has been reached, no noticeable decline in responsiveness occurs, even with advanced development. Confirming that fully developed salivary gland is oncogenically responsive to PV, are experiments in which tumors were induced in 30 to 40 day-old or fully adult mice by (a) giving a very large dose of virus intravenously (11); and (b) exposing adult mice to 300 r of total body irradiation before giving a standard dose of virus intravenously (12).

Concerning the mechanisms that operate to make an oncogenic response to PV possible during salivary gland development, only speculation is possible. Perhaps a specific gene locus is activated (involving change of DNA from a condensed to an extended state), enabling virus to become integrated in some key position. Such a change (gene activation) may be classed among epigenetic mechanisms, and since it is known that epithelio-mesenchymal interaction is essential for normal morphogenesis to occur, one supposes that this interaction is a *sine qua non* both for morphogenesis and for PV-induced oncogenesis. This would explain the failure of embryonic oral epithelium to give rise to tumors before morphogenesis had proceeded to a threshold level.

Experiments with isolated epithelial and mesenchymal components of submandibular rudiments have shed further light on this aspect of tumor genesis (7). When either component was infected and transplanted separately to new hosts, no tumors appeared. If, however, the two components were recombined after trypsinization and infection, then typical polyoma-type tumors arose in the transplants, showing that the trypsinization procedure itself did not abolish oncogenic responsiveness. When this experiment was performed in the *in vitro* system, the results were comparable (7).

Since these experiments were performed with components of rudiments that had already started to undergo morphogenesis, and in which conditioning of the genome had therefore theoretically already been established, some additional function of epithelio-mesenchymal interaction must be essential. Again, only speculation is possible, but our experience, like that of developmental biologists, has shown that the epithelium of submandibular rudiments is dependent on specific capsular mesenchyme for continued proliferation as well as morphogenesis. It seems reasonable that the same sort of dependency continues to exist during the early stages, at least, of oncogenesis induced by PV. This would explain the failure of neoplasms to appear from isolated, infected components of salivary gland rudiments. For a working hypothesis then, we have a concept in which salivary gland epithelium infected by PV is able to undergo certain changes leading to acquisition of neoplastic properties only if "helper cells" (capsular mesenchyme) are also present. This concept is not unique, as it is similar to the concept of dependency of certain tumors of endocrine target organs upon the appropriate hormone, e.g., thyroid tumors dependent on thyrotropin, as described by Furth and Clifton (13). The main difference is that in the PV-salivary gland system the neoplastic cells are dependent on an intimate association with presumably specific supporting cells, whereas in the endocrine-dependent systems the neoplastic cells depend upon a hormone that may be transported humorally from a distant site.

Recent Experiments

With this background information and a concept of the epithelio-mesenchymal complex as an integrated unit of neoplastic response, we decided to determine the effect of modifications of the epithelio-mesenchymal complex on the neoplastic response. Work of Grobstein (10) had shown that several mesenchymes other than that of submandibular capsule would not support morphogenesis. This made it particularly pertinent to learn whether "alien" mesenchymes in combination with submandibular epithelium would permit oncogenesis in response to PV.

To some extent, the substitution of alien mesenchyme had already been shown inadequate to support tumor genesis in those experiments in which infected, isolated submandibular epithelium was transplanted subcutaneously to syngeneic newborn mice (7), for the transplanted epithelium necessarily established contact with subcutaneous connective tissues. The latter apparently could not substitute for salivary mesenchyme in oncogenesis.

However, it seemed of more interest and meaning to recombine submandibular epithelium with mesenchyme from other organs in which epithelio-mesenchymal interactions are involved during morphogenetic inductions. Perhaps most meaningful would be reciprocal interchanges of epithelium and mesenchyme from 2 organs both of which were oncogenically responsive to PV. Conceivably, those organs oncogenically responsive to PV might have mesenchymes with some epigenetic factor common to all, and this could be demonstrated by interchange experiments. On the other hand, the specificity of mesenchyme for PV tumor-induction in the submandibular gland might be so absolute that no other mesenchyme could be substituted.

Experiments were at first designed to construct what may be called phenotypic chimeric crosses between rudimentary hair follicles and rudimentary submandibular glands. This particular cross proved to be unsuitable because, although it was shown that rudimentary hair follicles would develop and produce hairs during one prolonged anlagen phase after transplanting (14), none of the tumors characteristically induced by PV in native hair follicles ever appeared in the infected transplants (unpublished observations). Further, we have never succeeded in transplanting even the actively growing tumors induced in hair follicles by PV.

Therefore, we turned to tooth rudiments for a source of mesenchyme to be exchanged with that of submandibular rudiments. It had been shown that ameloblastic tumors are induced in the dental organ by PV (11, 15, 16), and Main and Dawe (17) found that such tumors appeared in 38 percent of PV-infected, 14-day mandibular incisor tooth germs transplanted to syngeneic newborn mice.

The details of the techniques used are in press elsewhere (7), so only the main features of this experiment will be given here. Submandibular salivary gland rudiments and mandibular incisor tooth rudiments were dissected from C3H/Bi mouse embryos on the 13th or 14th day of development. In separate groups, these were dissociated into epithelial and mesenchymal components with the aid of trypsin and gentle mechanical dissection. (See Figs. 1 to 6) Reciprocal recombinations of the components were then made, using an average of about 7 components of each type per cluster of recombinants (Table IV). The recombined clusters were then cultured for 24–28 hours to allow re-establishment of epithelio-mesenchymal contact and interaction. This was done in 1.5 ml. of medium in Grobstein dishes, incubated at 36° C in a partially humidified atmosphere of 5 percent CO_2 in air. The medium was Eagle's Minimal Essential Medium (18) supplemented with calf serum at 10 percent of total volume and containing 100 u/ml. of penicillin G and 0.002 percent phenol red. Two or three times during the culture period the re-approximation of epithelium and mesenchyme was assisted, where indicated, by pushing the components into contact with each other by means of cataract knives.

At the end of the period of recombination in culture, three-fourths of the medium was replaced by a suspension of polyoma virus, in which the recombined clusters remained for 2 hours. The clusters were then transplanted to newborn C3H/Bi mice bred in our polyoma-free breeding colony. The particular pool of virus used was 2-PTA3, representing the third culture passage of a strain isolated in this laboratory from a salivary gland tumor. This pool induced salivary gland tumors in 22 of

32 (69 percent) C3H/Bi mice that received 0.05 ml. subcutaneously within 48 hours after birth.

Controls consisted of intact salivary gland rudiments and dental rudiments, also cultured for 24–28 hours and infected with the same pool of virus in the same manner as the recombined components prior to transplantation. Fewer rudiments per recipient were transplanted in the control groups than in the recombined groups, however. (See third vertical column in Table IV.)

To determine the results of epithelio-mesenchymal interactions in the absence of virus, the same procedures outlined above were followed, but infection with PV was omitted. The recipients were killed 15 to 21 days after receiving the grafts, and those grafts that could be found, either grossly or with the aid of a dissecting microscope, were examined histologically in serial sections.

Animals that received PV-infected intact rudiments or either of the component recombinations were checked weekly for appearance of tumors, either at the site of transplant or in native organs of the host. At this writing, the animals have been followed for periods ranging from 7 to 10 months. When death appeared imminent, tumor-bearing animals were killed and detailed necropsies were performed. Seventy-three of the 125 animals listed in Table IV had undergone necropsy at the time the data were tabulated.

Results

A. Intact rudiments and rudiment crosses infected with virus. No epithelial tumors of any type were found in the infected transplants of salivary epithelium recombined with dental mesenchyme (horizontal column headed SE × DM in Table IV), or in the reciprocal recombinations (DE × SM). Nor did tumors appear in the infected transplants of intact dental rudiments. As was to be expected from previous experience, typical polyoma-type salivary gland tumors arose in transplants in 17 of 27 recipients, and salivary gland tumors were identified in native glands of 42 to 70 percent of the recipients in all 5 groups in Table IV. Therefore there is no question that the viral preparation had adequate oncogenic activity for salivary gland. Apparently it had little oncogenic activity for dental organ, however, as no neoplasms appeared in the transplanted incisor rudiments. Whether this characteristic of the virus preparation accounts entirely for the failure of tumor induction in SE × DM and DE × SM transplants cannot be estimated, but in the absence of any positive result with substitution of dental for salivary mesenchyme, we failed to produce evidence invalidating the concept that the interaction of salivary mesenchyme is essential in potentiating oncogenic responsiveness to PV. The experiment has also yielded no evidence that would support the concept that one or the other of the components of the epithelio-mesenchymal complex dominates in determining oncogenic responsiveness. If, for example, salivary mesenchyme were the controlling factor in making the epithelio-mesenchymal complex responsive, then some tumors should have appeared in the DE × SM combinations. If salivary epithelium played the dominant part, some tumors should have appeared in the SE × DM combinations. Neither of these possibilities occurred, and we are left with the presumption that a correct match, or some specific complementarity of epithelium and mesenchyme, is a critical factor in making the oncogenic response possible.

Not shown in Table IV is the fact that in 3 recipients of the DE × SM combinations and in 2 recipients of the SE × DM combinations, fibrosarcomas appeared in or near the areas where transplants were deposited. These were histologically identical with the fibrosarcomas frequently induced by PV in the subcutaneous tissues, and because of this and the absence of any association with a recognizable transplant, these tumors were presumed to have arisen from host connective tissue. The possibility cannot be dismissed that they originated from transplanted salivary or dental mesenchyme, but the lack of any special histologic features gave no support to this supposition.

B. Intact rudiments and rudiment crosses not infected with virus. Study of the extent and form of development taken by SE × DM and DE × SM combinations provided information that is relevant in explaining the results recorded in Table IV.

It was found (Table V) that, with certain exceptions, there was very little organoid development from either of the two unnatural combinations (Figs. 7–9). In only one instance was even slight evidence of tooth development found (Fig. 8), and in the majority of recovered chimeric combinations, only small epithelium-lined cysts were found in serial sections. As one might expect from knowledge of the faculties of dental and salivary epithelium, under normal and pathologic conditions, most of the cysts from the SE × DM combination were lined with columnar epithelium, and most of those from the DE × SM combination were lined with squamous epithelium. This probably signifies that the epithelium of the respective

TABLE IV

Absence of Tumor Development in Reciprocal Recombinations of Epithelium and Mesenchyme from Salivary Rudiments and Dental Rudiments, Transplanted after Infection with Polyoma Virus

Rudiment or Component Crosses	Total No. Rudiments Used	Av. No. Ruds. per Recipient	Tumors in Transplants		Tumors in Hosts	
			Fraction	Percent	Fraction	Percent
SE × DM	262 × 276	7.3 × 7.5	0/36	0	22/36	61
DE × SM	257 × 242	7.1 × 6.7	0/36	0	25/36	70
SE × SM *	188 × 188	4.0 × 4.0	11/47	23	29/47	62
Sal. Ruds.	62	2.3	17/27	63	18/27	67
Dent. Ruds.	54	2.0	0/26	0	11/26	42

SE = Submandibular rudiment epithelium.
SM = Submandibular rudiment mesenchyme.
DE = Dental rudiment epithelium.
DM = Dental rudiment mesenchyme.

* Data in this horizontal column taken from results published in reference (7).

TABLE V

Findings in Uninfected Grafts of Reciprocal Recombinations of Epithelium and Mesenchyme from Salivary and Dental Rudiments

Material Transplanted	No. of Ruds.	No. of Recips.	Time in Host	Graft Recov.	Cysts Columnar Lined	Cysts Squamous Lined	Bone	Developing Sal. Gland	Developing Tooth
Intact Sal. Rud.	28	8	15 & 20 d.	4	1	0	0	3 (good)	0
Intact Dent. Rud.	14	4	15 & 21 d.	3	0	3	3	0	2 (good)
SE × DM	54SE × 47DM	8	15 & 20 d.	5	4	2	4	3 (very slight)	0
DE × SM	47DE × 54SM	8	15 & 20 d.	7	1	4	1	2 (good)	1 (very slight)

SE = Submandibular rudiment epithelium.
SM = Submandibular rudiment mesenchyme.
DE = Dental rudiment epithelium.
DM = Dental rudiment mesenchyme.

rudiment types could not be redirected by interaction with mesenchyme of the other organ.

The presence of bone in 4 of 5 SE × DM combinations is explained on technical grounds. The mandibular incisor rudiment lies in close apposition to mandibular mesenchyme with ultimate bone-forming potential, and in all probability some of this material (Meckel's cartilage and pre-cartilage) was included with the specific dental mesenchyme during the dissections. The bone found in 1 of the DE × SM combinations is more difficult to explain, though it could have resulted either from metaplastic bone formation in the host or from inadvertent contamination of the dental epithelial components by some of the mandibular cartilage or pre-cartilage thought to be the source of bone in SE × DM combinations.

The finding of advanced salivary gland morphogenesis in 2 transplants of the DE × SM combination appears to be an exception to the general rule that no development occurred in crossed rudiments. Although we cannot be certain at this time, there is a probable explanation for this, again technical in nature. In Fig. 5 it can be noted that in isolating dental epithelium, we commonly included a small rim of adjacent oral epithelium along with the bud of enamel-organ epithelium. It seems likely that this rim of oral epithelium could be responsive to salivary mesenchyme, inasmuch as minor salivary glands normally are found widely distributed beneath the oral mucosa.

Actually, of greater significance may be the observation of very slight morphogenesis in 3 of 5 SE × DM combinations. This observation is not likely the result of any technical fault, and may indicate some degree of capability of dental mesenchyme to support morphogenesis in epithelium from salivary gland. However, the extent of morphogenesis was very limited (Fig. 7). Epithelium of other organs (thymus, pancreas) is known to undergo morphogenetic development under the influence of unnatural mesenchymes (19, 20).

Discussion

Even though the recent segment of work presented in Tables IV and V entailed the use and observation of more than 800 salivary gland and dental rudiments, certain features in the results suggest that it may be profitable to extend the experiments, with certain modifications, before judging whether or not it is possible to obtain neoplasms from rudiments with substituted mesenchymes.

For example, it seems desirable to repeat the experiments, using the LID strain of polyoma virus, which has been shown much more effective in inducing odontogenic tumors than the strain used here (16, 17). In the present work, tumors failed to appear even in intact, transplanted dental rudiments, and this suggests that any combination containing some component of dental rudiment might have a lowered degree of oncogenic responsiveness to this particular virus.

Further work should also be carried out to determine with more certainty just what the developmental capabilities of SE × DM and DE × SM combinations are. If, for example, it can be shown that salivary epithelium can be induced by dental mesenchyme to undergo some moderate degree of morphogenesis, it would not seem entirely futile to continue attempts to induce tumors in this combination.

It must be acknowledged that in planning these experiments, we anticipated that some tumors might emerge from the artificial combinations tested. This anticipation was based on the foreknowledge that both the salivary gland rudiments and the dental rudiments, when left intact, can give rise to tumors in response to PV. Because the tumors in these two organs are histologically distinct (e.g., the odontogenic tumors produce keratin while the salivary tumors do not), the opportunity seemed ideal for determining whether certain characteristics of each tumor type might be imparted by the epithelial component, or by the mesenchyme. In short, the system seemed right for analysis of epigenetic influences on the determination of "unit characters" (21) of neoplasms.

This system differed from that used by Wang (1) in analysis of the controlling factors in feather formation, in that the reciprocal exchanges of mesenchyme in our work were between different organ types, whereas in Wang's experiments the

---

Fig. 1. Intact submandibular salivary gland rudiments (Stages 2 and 3) from 13-day mouse embryos, freshly dissected from the donors. Central light areas are epithelium, dark peripheral zones are mesenchyme. ×34.

Fig. 2. Intact mandibular incisor tooth rudiments, dissected from the same embryos that served as donors of the salivary gland rudiments shown in Fig. 1. Epithelium is seen as light, u-shaped ingrowths (profile) or irregular rings (dorso-ventral view). Mesenchyme is dark, peripheral material. ×34.

Fig. 3. Epithelial components separated by trypsinization from the salivary gland rudiments shown in Fig. 1. Note stalks (early ducts) and bulbs, which subdivide during morphogenesis and give rise to adenomeres. ×34.

Fig. 4. Mesenchymal components separated by trypsinization from the salivary gland rudiments shown in Fig. 1. ×34.

Fig. 5. Epithelial components separated by trypsinization from the tooth rudiments shown in Fig. 2. Note that in some of these preparations a rim of contiguous oral epithelium is still attached to the knob-like enamel organ, which has not yet been invaginated by dental papilla. ×34.

Fig. 6. Mesenchymal components separated by trypsinization from the tooth rudiments shown in Fig. 2. They are very similar in appearance to the mesenchyme of salivary rudiments (Fig. 4). However, in handling they are less sticky and adhere less readily to epithelium of either salivary gland or dental origin. ×34.

Fig. 7. Transplant of salivary epithelium combined with dental mesenchyme, 20 days after implantation in newborn host. There is slight organization of the epithelium into a gland-like structure, but this represents only minimal morphogenesis. Note that salivary epithelium lies in a groove in bone which has developed probably from a bit of Meckel's cartilage included with dental mesenchyme. ×125.

Fig. 8. Transplant of dental epithelium combined with salivary mesenchyme, 15 days after implantation in newborn host. Most of the epithelium forms the wall of an epidermoid cyst, but at lower right a projection of non-keratinizing epithelium resembles early enamel organ. This represents very slight morphogenesis, at best. ×85.

Fig. 9. Three cystic structures recovered from the site of transplant of PV-infected combination of dental epithelium and salivary gland mesenchyme, 4 months after the transfer. The linings of the cysts are composed of flattened or cuboidal epithelium, 2 to 3 cells deep and in places heaped up in small papillae. The lumens contain exfoliated cells and fibrillar material which is not birefringent, hence not clearly keratinaceous. Note lymphocytic aggregates (dark areas) adjacent to the grafts. The host bore tumors in its own salivary glands. ×95.

Fig. 10. Section of tumor that appeared in a PV-infected graft of salivary gland epithelium recombined with salivary gland mesenchyme, 2 months after implantation in the newborn host. This histologic pattern, with a mixture of acinar and solid components, is typical of PV-induced tumors in intact salivary glands. ×200.

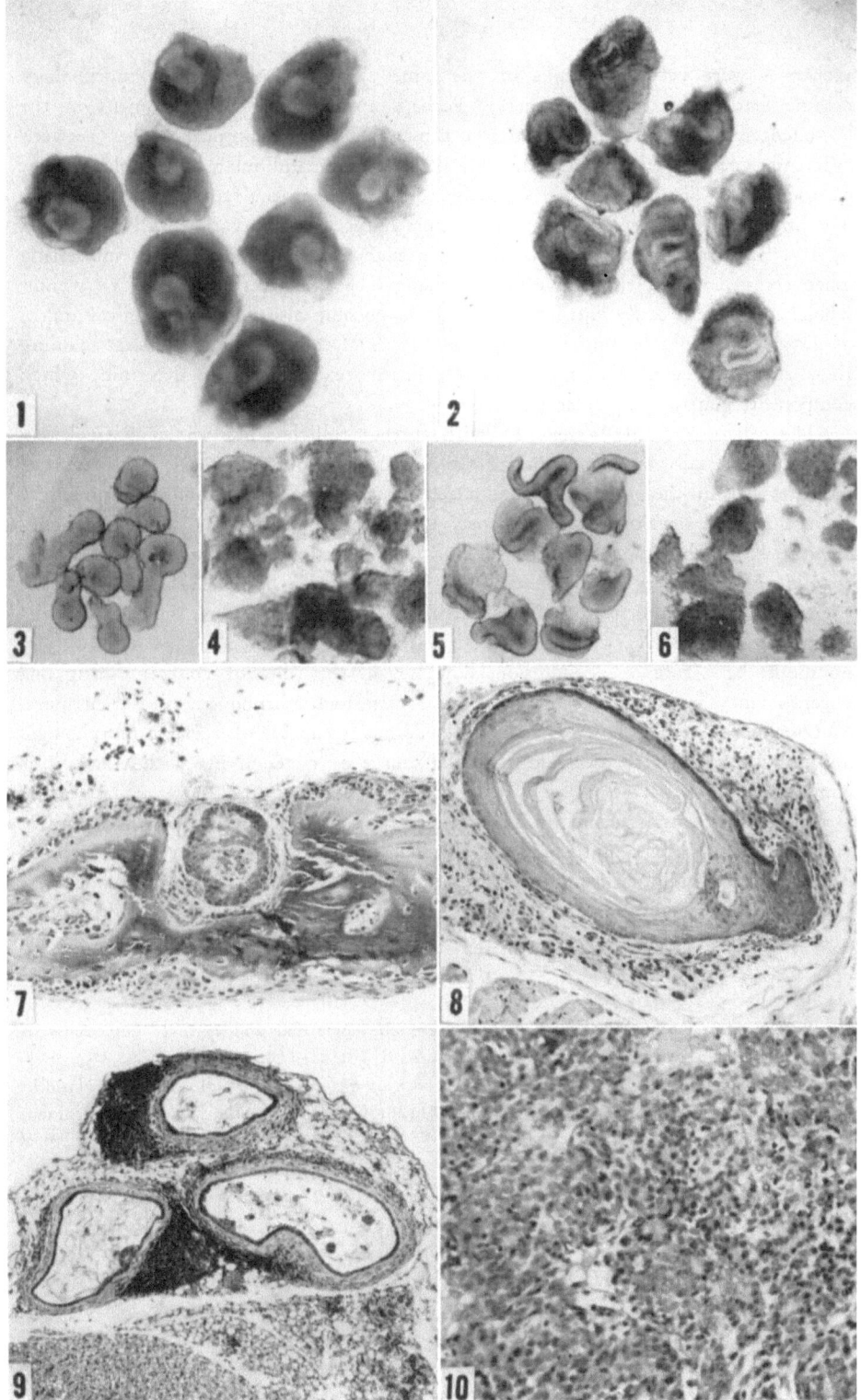

31

exchanges were between organs of the same type, but of different morphology (saddle feathers vs. breast feathers). Perhaps a more rational experiment in the PV-oncogenic system would be to test the responses of reciprocal crosses between different types of salivary gland, e.g., between submandibular and sublingual or parotid glands. The disadvantage of such exchanges lies in the fact that histologically it is impossible to differentiate among tumors induced in the various salivary glands by PV. This fact nevertheless would make such experiments especially interesting when compared with those involving exchanges between teeth and salivary glands, whose tumors are easily distinguishable. It is possible also that biochemical marks of distinction might be found among the 3 types of salivary gland tumors, enabling these glands to be used in mesenchyme exchange experiments to determine which components control particular characters.

The experiments of Wang (1) with feather follicles, like those of Grobstein (9, 10) with salivary gland and of Golosow and Grobstein (20) with pancreas, indicate that in these secondary induction systems, the epithelium determines the pattern of the developmental response, while the mesenchymal component only determines whether morphogenesis will or will not occur. In the case of feathers, the line of symmetry is also established by the mesenchyme. These observations suggest that if tumors arose from the SE × DM crosses, they should have the characteristics of salivary tumors, rather than of odontogenic tumors. This can by no means be a foregone conclusion, however, as the rules of control during oncogenesis may not be the same as those for normal morphogenesis. Furthermore, McLoughlin (22) has shown that chick epidermis is capable of variable morphologic and functional activity, depending on the source of mesenchyme with which it is combined.

At any rate, the failure of any tumors to arise from either of the chimeric rudiment types obliges us to hold for the moment with the concept that, for cells of mouse submandibular rudiments, the oncogenic response to PV is dependent on sustained and specific interaction between salivary gland epithelium and salivary gland mesenchyme.

References

1. WANG, H.: The morphogenetic functions of the epidermal and dermal components of the papilla in feather regeneration. *Physio. Zool.* 16:325–350, 1943.
2. DAWE, C. J.: *Cell sensitivity and specificity of response to polyoma virus.* National Cancer Institute Monograph No. 4, pp. 67–128. Washington, D. C., U. S. Government Printing Office, 1960.
3. ARISTOTLE: On the generation of animals. See especially Book II, Chapter 1, Berlin No. 734ª, translated by D'Arcy Wentworth Thompson. In *The Works of Aristotle.* Chicago, Encyclopaedia Britannica, Inc., 1952, vol. 2.
4. OPPENHEIMER, JANE M.: Problems, Concepts and Their History. In *Analysis of Development.* Willier, B. H., Weiss, P. A., and Hamburger, V. (Eds.), Philadelphia, W. B. Saunders Company, 1955, pp. 1–24.
5. DAWE, C. J., and LAW, L. W.: Morphologic changes in salivary-gland tissue of the newborn mouse exposed to parotid tumor agent *in vitro. J. Nat. Cancer Inst.* 23:1157–1177, 1959.

6. DAWE, C. J., LAW, L. W., MORGAN, W. D., and SHAW, M. G.: Morphologic responses to tumor viruses. *Fed. Proc.* **21**:5–14, 1962.

7. DAWE, C. J., MORGAN, W. D., and SLATICK, M. S.: Influence of epithelio-mesenchymal interactions on tumor induction by polyoma virus. *Int. J. Cancer. In press.*

8. MOSCONA, A.: Cell suspensions from organ rudiments of chick embryos. *Exp. Cell Res.* **3**:535–539, 1952.

9. GROBSTEIN, C.: Inductive epithelio-mesenchymal interaction in cultured organ rudiments. *Science.* **119**:52–55, 1953.

10. GROBSTEIN, C.: Epithelio-mesenchymal specificity in the morphogenesis of mouse submandibular rudiments *in vitro. J. Exptl. Zool.* **124**:383–414, 1953.

11. DAWE, C. J., LAW, L. W., and DUNN, T. B.: Studies of parotid-tumor agent in cultures of leukemic tissues of mice. *J. Nat. Cancer Inst.* **23**:717–797, 1959.

12. LAW, L. W., and DAWE, C. J.: Influence of total body x-irradiation on tumor induction by parotid tumor agent in adult mice. *Proc. Soc. Exp. Biol. Med.* **105**:414–419, 1960.

13. FURTH, J., and CLIFTON, K. H.: Experimental pituitary tumors and the role of pituitary hormones in tumorigenesis of the breast and thyroid. *Cancer.* **10**:842–853, 1957.

14. FRIEDMAN-KIEN, A. E., DAWE, C. J., and VAN SCOTT, E. J.: Hair growth cycle in subcutaneous implants of skin. *J. Invest. Dermatology.* **43**:445–450, 1964.

15. STANLEY, H. R., DAWE, C. J., and LAW, L. W.: Oral tumors induced by polyoma virus in mice. *Oral Surg.* **17**:547–558, 1964.

16. STANLEY, H. R., BAER, P. N., and KILHAM, L.: Oral tissue alterations in mice inoculated with the Rowe substrain of polyoma virus. *Periodontics.* **3**:178–183, 1965.

17. MAIN, J. H. P., and DAWE, C. J.: Tumor induction in transplanted tooth buds infected with polyoma virus. *J. Nat. Cancer Inst. In press.*

18. EAGLE, H.: Amino acid metabolism in mammalian cell cultures. *Science.* **30**:432–437, 1959.

19. AUERBACH, R.: Morphogenetic interactions in the development of the mouse thymus gland. *Develop. Biol.* **2**:271–284, 1960.

20. GOLOSOW, N., and GROBSTEIN, C.: Epithelio-mesenchymal interaction in pancreatic morphogenesis. *Develop. Biol.* **4**:242–255, 1962.

21. FOULDS, L.: Some problems of differentiation and integration in neoplasia. In *Biological Organization at the Cellular and Supercellular Level.* Harris, R. J. C. (Ed.), New York, Academic Press, 1963, pp. 229–244.

22. McLOUGHLIN, C. B.: The importance of mesenchymal factors in the differentiation of chick epidermis. II. Modification of epidermal differentiation by contact with different types of mesenchyme. *J. Embryol. Exp. Morphol.* **9**:385–409, 1961.

The Response of Metanephric Rudiments to Polyoma Virus *in Vitro* [1]

Discussion

BY W. H. KIRSTEN [2] AND T. P. WEIS

Department of Pathology, Division of Biological Sciences,
The University of Chicago, Chicago, Illinois

Dr. Dawe's observations on the response of organ cultures of mouse submandibular rudiments to polyoma virus (PV) can be summarized as follows: (a) the basic unit of the neoplastic response is the epithelio-mesenchymal complex, (b) neoplasms fail to appear from the isolated epithelial or mesenchymal components of the salivary-gland rudiments and (c) acquisition of neoplastic properties by salivary-gland epithelium requires "helper cells" which are provided for by salivary capsular mesenchyme but not by odontogenic mesenchyme. Furthermore, the responsiveness to PV oncogenesis was not maintained by exposure of "chimeric" dental epithelium and salivary-gland mesenchyme.

I would like to extend this information by presenting some experiments with a different, though comparable host system, namely, the response of metanephric rudiments to PV in organ culture. The experiments are based on the findings by Grobstein, et al. (3–6), who have shown that the 11-day-old mouse metanephric rudiment undergoes a characteristic morphogenesis in vitro. The epithelial and mesenchymal components interact to form tubules. Developmental interdependency between the two components is indicated by the inability of either the ureteric bud epithelium or the mesenchyme to continue tubular morphogenesis when grown in isolation. Tubules are produced by the metanephrogenic mesenchyme which interacts with the epithelial component. The latter can be substituted for by embryonic submandibular salivary epithelium or by dorsal spinal cord, but not by a number of other living embryonic or adult tissues. The tubule-inducing effect of metanephrogenic mesenchyme or embryonic mouse spinal cord can be transmitted across a filter membrane of 20 to 30 microns thickness which prevent the exchange of cells between the inductively active tissues.

We have used metanephric rudiments from embryonal mice and rats to study whether or not malignant transformation by PV can be induced in organ cultures and, if so, to explore the role of the epithelial and mesenchymal components in the morphogenesis of such tumors. The technical details have been described elsewhere (9). The results can be summarized as follows:

The response of metanephric rudiments to PV oncogenesis in vitro was species specific. Malignant transformation did not occur in mouse metanephric rudiments from 11-day or 14-day-old (C_3H/fAn $\times$ BALB/cJ)F_1 embryos infected with PV one day after the organ cultures were established. Although tubular differentiation continued for a few days (Fig. 1),

[1] This investigation was supported by grant C-4311 from the National Cancer Institute, National Institutes of Health, Public Health Service.
[2] USPHS Career Development Awardee.

cytopathic changes in the ureteric bud epithelium and in the mesenchymal cells became apparent as early as 5 days after infection (7–9). The entire organ cultures eventually became necrotic and cellular growth was completely arrested. Attempts to rescue virus-infected organ rudiments by transfer from the millipore filter assembly to a gelatine sponge to provide a more favorable matrix for the diffusion of nutrient media, failed to maintain viability of the organ cultures. No proliferative response has been observed with mouse rudiments in over 40 different combinations of mesenchyme and bud epithelium either grown together or separated by a thin filter membrane or dissociated by trypsin and reaggregated by rotation (9). We concluded that PV affected mouse metanephric cells in a manner similar to the cytopathic effects seen in cell cultures of mouse fibroblasts.

In contrast to mice, metanephric organ cultures from embryonal rats responded to PV infection by cell proliferation rather than lysis. Intact metanephric rudiments from 11-day rat embryos of the inbred Wistar/Furth (W/Fu) (2) strain were exposed to the LID-1 strain of PV. Control cultures consisted of noninfected metanephric organ cultures obtained from littermate embryos that served as donors for the virus infected cultures. The total mass of the PV-infected metanephric rudiment exceeded the size of the uninfected organ cultures after 10 days in vitro. A first possible indication of neoplastic properties acquired by PV-exposed cultures was noted when infected and noninfected rudiments were transferred from the filter assembly to a gelatine sponge at 3 weeks after the cultures were initiated. Growth of noninfected cultures occurred predominantly at the surface of the sponge or extended into the upper one-third of the sponge meshes, while fibroblast-like cells from virus-infected rudiments occupied the entire thickness of the sponge. Tubules were clearly recognizable in both types of culture at the sponge surface. Whether or not "sponge invasion" represented a valid criterion for malignant transformation has not been determined. The cultures were terminated after 24 to 28 days in vitro. Examination of PV-infected rudiments showed many tubular structures embedded in a primitive mesenchyme (Fig. 2). Between 4 to 40 cultures from the experimental or control groups were pooled and transplanted by the trocar method either subcutaneously into the fat pad in a previously marked interscapular area or between the peritoneum and the fascia of the upper abdominal muscles. Recipients were weanling inbred W/Fu rats whose serum had been tested for polyoma-virus antibodies by hemaggluti-nation-inhibition tests (2). Only polyoma-free rats were selected as recipients in order to avoid the transplantation rejection phenomenon (Habel, this symposium). None of 17 trans-plants from noninfected control cultures was recoverable from the transplantation sites after 3 to 12 weeks of observation. Microscopic examination revealed fibrosis or non-specific granulation tissue at the sites of the grafts. In contrast, virus-infected metanephric rudi-ments formed palpable tumor nodules at the site of implantation within 2 to 3 weeks. The tumors attained large sizes and eventually ulcerated through the skin. Careful histologic examination of the remaining host tissues failed to reveal distant metastases, although the transplant infiltrated the surrounding muscle. Biopsies from the tumor grafts showed the typical primitive tubular renal sarcoma described previously (2). Continuous growth and locally invasive properties acquired by grafts of PV-infected cultures as opposed to non-infected cultures was taken as evidence for neoplastic conversion. The lack of distant metas-tases may indicate that certain "unit characters" of the neoplastic potential had not been expressed in the sequence of steps to fully autonomous neoplasia.

The role of the mesenchymal and epithelial components in transformation by PV was studied further. First, metanephric rudiments from 11-day W/Fu rat embryos were separated by trypsinization and gentle agitation into the mesenchymal cap and ureteric bud epithelium (4, 5). The two components were grown in the same culture dish but separated from another by a millipore filter membrane 45 microns thick. Each organ rudiment was infected

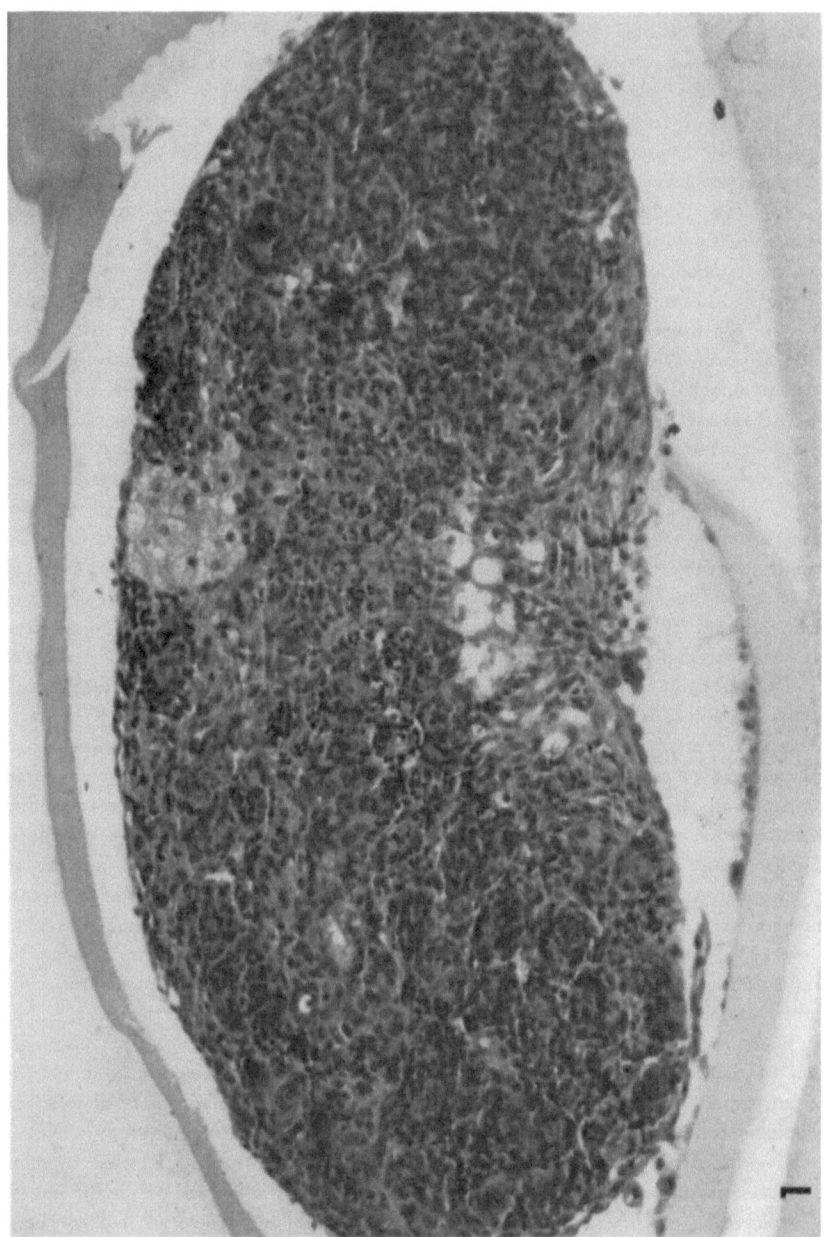

Fig. 1. Mouse metanephric rudiment 10 days after PV infection. There is considerable necrosis and beginning disintegration of the organ rudiment. Magnification 51.2$\times$

Fig. 2. Complete rat metanephric rudiment 16 days after PV infection. The organ culture is viable, tubulogenesis continues. Magnification 51.2×

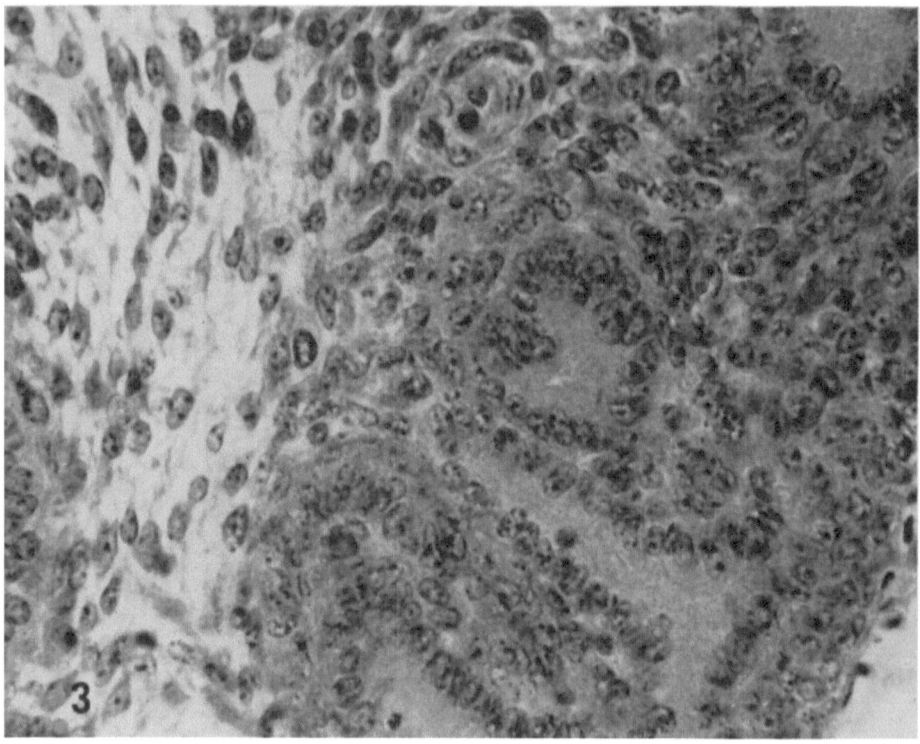

Fig. 3. Tubulogenesis in the mesenchyme of the separated kidney rudiment. The components were grown in the same culture dish separated by a filter membrane. Magnification 51.2✕

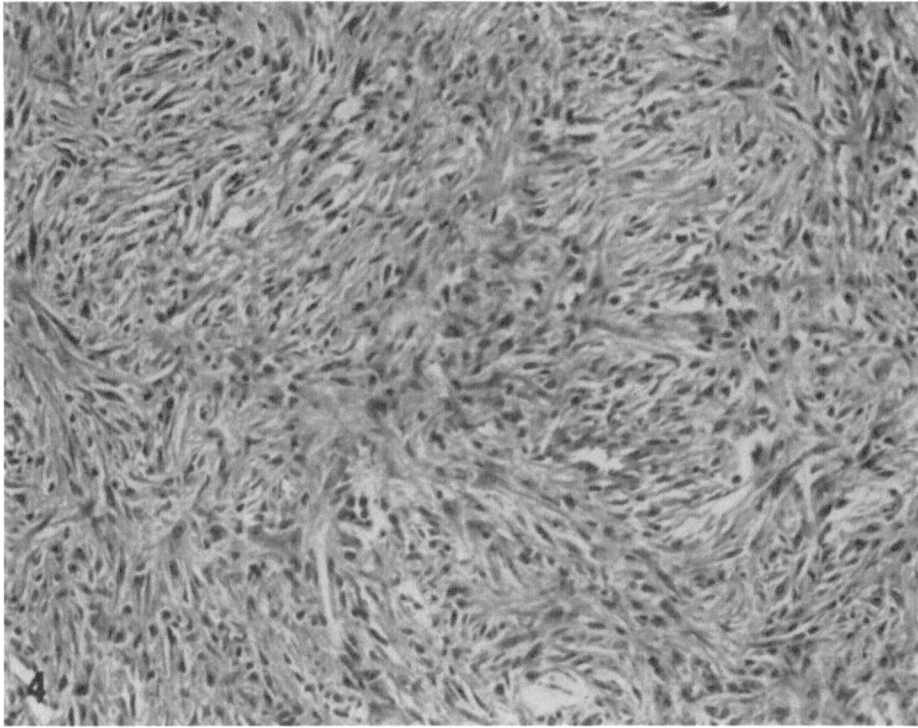

Fig. 4. Primitive sarcoma without tubule formation induced by PV in vitro from the metanephric mesenchyme. Magnification 51.2✕

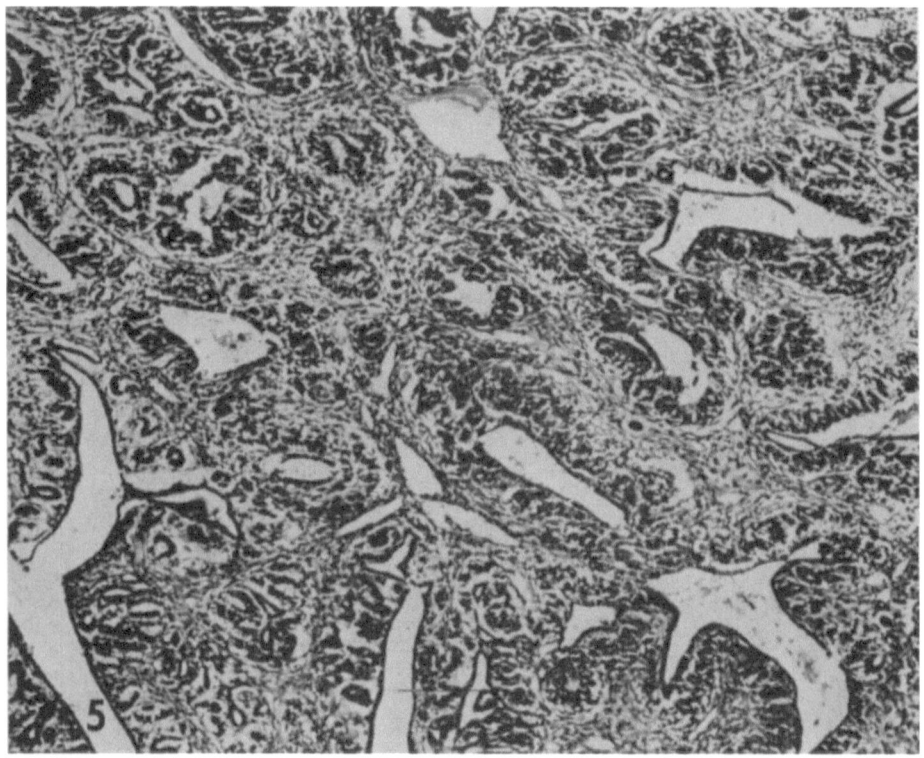

Fig. 5. Transplant of metanephric rudiment transformed by PV. Many tubules and some preglomeruli are embedded in a primitive mesenchymal stroma. Magnification 20.5×

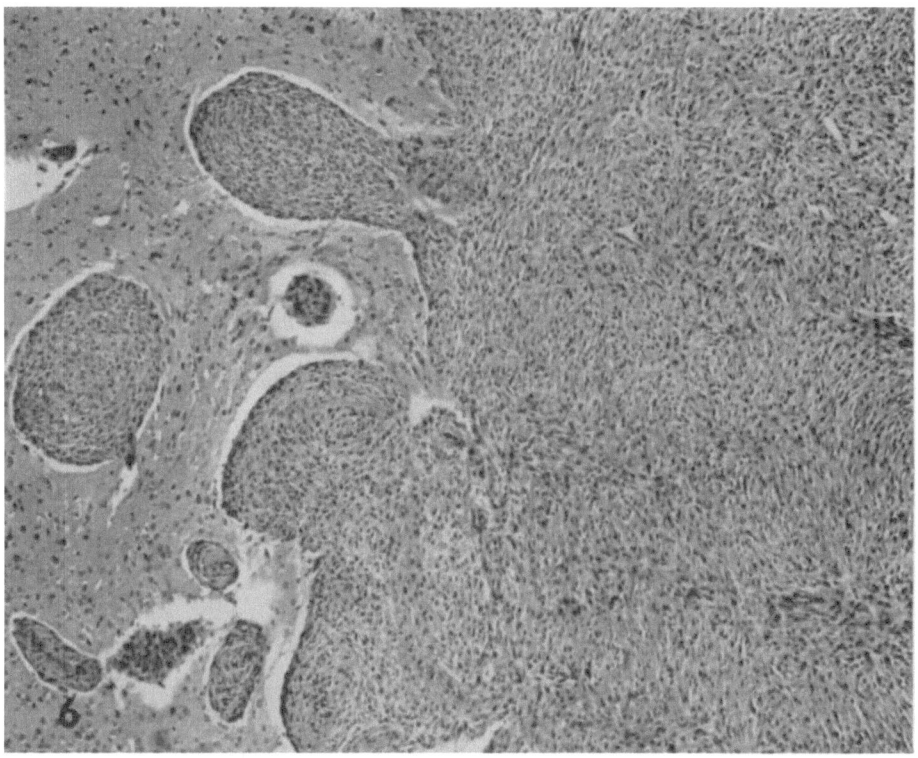

Fig. 6. Intracerebral transplant of the mesenchymal component transformed by PV. Magnification 20.5×

with PV. After 24 to 27 days in vitro they were then transplanted subcutaneously or intraperitoneally to W/Fu rats as pools of 5 to 6 mesenchymal caps or as pooled ureteric bud components. The separation of the whole rudiments into mesenchyme and epithelium did not affect their transforming ability as judged by continuous growth of all 12 combinations from different embryo litters tested thus far. Morphologic evaluation of the mesenchymal and epithelial components before and after transplantation revealed primitive embryonal nephromas with abundant tortuous or straight tubules, occasional preglomeruli and primitive mesenchymal stroma (Figs. 3, 5). Thus, tubulogenesis was maintained and even exceeded the number of tubules seen in tumors induced in vivo or by exposure of the intact metanephric rudiment to PV.

Finally, the metanephric rudiments from 11-day-old W/Fu rat embryos were dissociated into mesenchyme and ureteric bud. Each component was grown in a separate tissue-culture dish for periods of 3 to 4 weeks. The results were highly variable when the isolated epithelial component was exposed to PV. Only 1 of 24 epithelial rudiments remained viable for 3 weeks or more under the culture conditions used. Several attempts to transplant these rudiments (after 14 and 18 days in vitro) did not yield discernible tumors grossly nor could the grafts be recovered from the implantation sites. Further experiments are needed before the response of the isolated epithelium can be fully evaluated. However, the isolated mesenchymal cap continued to grow following PV infection, although tubule formation was not observed by inspection of the cultures or by histologic examination after 28 days in vitro (Fig. 4). Moreover, the presumably transformed mesenchymal caps were transplantable only to such immunologically privileged sites as the central nervous system or testis (Fig. 6). The growth rate of these tumors was slow, and distant metastases were not observed. Although the mesenchymal component of the rat metanephric rudiment is transformable into an invasively growing tumor, the capacity of the mesenchyme to produce tubules is lost in the absence of the epithelial rudiment.

The difference in the morphogenetic response to PV observed in Dr. Dawe's and our experiments can be explained by a consideration of the embryogenesis of the two systems. Salivary gland rudiments consist of an ectodermal and a mesodermal component whereas metanephric mesenchyme and ureteric bud epithelium are both mesodermal derivatives. As predicted by Dr. Dawe, "the renal site happens to be an opportune one for investigation of possible relationships between morphogenetic influences, viral susceptibility, and type of response" (1).

References

1. Dawe, C. J.: Cell Sensitivity and Specificity of Response to Polyoma Virus. Natl. Cancer Inst. Monogr. 4, 67–128 (1960).
2. Flocks, J. S., Weis, T. P., Kleinman, D. C., and Kirsten, W. H.: Dose-Response Studies to Polyoma Virus in Rats. J. Nat. Cancer Inst. 35, 259–284 (1965).
3. Grobstein, C.: Inductive Interaction in the Development of the Mouse Metanephros. J. Exp. Zool. 130, 319–340 (1955).
4. Grobstein, C.: Trans-Filter Induction of Tubules in Mouse Metanephrogenic Mesenchyme. Exp. Cell Res. 10, 424–440 (1956).
5. Grobstein, C.: Some Transmission Characteristics of the Tubule-Inducing Influence of Mouse Metanephric Mesenchyme. Exp. Cell Res. 13, 575–587 (1957).
6. Grobstein, C., and Parker, G.: Epithelial Tubule Formation by Mouse Metanephrogenic Mesenchyme Transplanted in Vivo. J. Natl. Cancer Inst. 20, 107–119 (1958).

7. SAXÉN, L., VAINIO, T., and TOIVONEN, S.: Effect of Polyoma Virus on Mouse Kidney
   Rudiment *in Vitro*. J. Natl. Cancer Inst. **29**, 597–631 (1962).

8. VAINIO, T., SAXÉN, L., and TOIVONEN, S.: Acquisition of Cellular Resistance to Polyoma
   Virus During Embryonic Differentiation. Virology **20**, 380–385 (1963).

9. WEIS, T. P., and KIRSTEN, W. H.: Interactions of Polyoma Virus and Organ Cultures of
   Embryonic Mouse Metanephric Rudiments. Arch. Path. **74**, 380–386 (1962).

Cell Transformation by the Mammary Tumor Virus *in Vitro* [1]

By

ETIENNE Y. LASFARGUES AND DAN H. MOORE

Department of Microbiology, Columbia University and the Rockefeller University
New York, New York

Introduction

One of the fundamental characteristics of the mammary cell is its changing physiology. A comprehensive study of virus-induced cell alterations must therefore take into account the intimate biological modifications which normally arise in the individual cell as well as in its immediate environment. The amount of information concerning the hormonal requirements for a normal development of mouse mammary glands is rather impressive but only a brief summary will be necessary to clarify the data presented in this report.

Both *in vivo* (1, 5, 10, 11, 12) and *in vitro* (3, 6, 15) studies concur to the existence of three active stages or phases in the progressive differentiation of the gland; they are followed by a resting, inactive period during which a reduced metabolism is maintained. The first stage begins with the advent of pregnancy: the mammary tissues, primarily under the influence of the ovarian and anterior-pituitary hormones grow rapidly, the epithelial cells forming the lobulo-acinar system of the mature gland. Around the 10th day of pregnancy the mitotic activity subsides and the combined effect of mammotropin and of the adrenal corticoids initiates a pre-secretory state. In this second phase, the epithelial cells became flatter and the enlarged lumen of the acini contains an accumulation of proteinaceous and fatty material. Full secretory activity is achieved after parturition mainly under the influence of the adreno-cortical hormones. This is the third, and final, stage of development. The resting gland itself is principally formed by a network of ducts whose end-bud has an active growth potential but remain dormant until the stimulation of the next pregnancy. Few, if any, viable acinar structures can then be found in the glandular mass.

In a series of previous experiments (7) mammary fragments from the agent-positive RIII strain were explanted in organ cultures at each one of their stages of development and confronted with the various hormones of the pregnancy cycle. Systematic electron-microscope scanning of the cultures showed that significant fluctuations in the elaboration of viral B particles by the epithelial cells could be induced. The B particles, now widely recognized as an image of the mammary

[1] Supported by grant CA-04588-07 from the National Cancer Institute, National Institutes of Health, United States Public Health Service.

44

TABLE I

PRODUCTION OF B PARTICLES IN AGENT-POSITIVE RIII
MAMMARY GLANDS, GROWN AS ORGAN CULTURES
OF THE THIRD ANTERIOR GLAND

Hormonal Stimulations	Early Pregnancy	Late Pregnancy	Lactation	Resting Gland
None	0	0	+	0
Ovarian Hypoph.	+++	0	+	+
Adreno-Cortical	0	0	+++	0

+: few, not budding B particles.
+++: large amount of B particles and active budding.

tumor virus (MTV), were produced in abundance in mammary fragments of early pregnancy and lactation only if placed under the proper hormonal stimulation (Table I). They were inhibited in the presence of hormones not compatible with the physiological state of the fragments. It was clear, therefore, that the synthesis of the milk agent is dependent on a specific state of the mammary cell and suggests that there exist periods of latency during which the virus has no activity.

Considering our problem of transformation of the mammary cell, it is difficult to conceive that a deep modification orienting the cell mechanisms towards malignancy and uncontrolled growth should be initiated by a virus in a state of latent parasitism. However, this could happen at a time of maximum production of the virus which, in this particular case, leaves two alternatives: early pregnancy or lactation.

Irreversibility and Differentiation

The complete differentiation into a mature mammary cell is irreversible and neoplastic transformation is not likely to occur in the lactating state. The possibility to explant fragments of mammary gland in various stages of differentiation and to place them in environmental conditions other than the one which prevail at the time of explantation was particularly informative on that subject (8). Apparently, when the mammary cell starts to differentiate under the stimulation of the hormones of early pregnancy, it progresses up to a point where reversion to a previous physiological situation becomes impossible. This seems to occur around the middle of pregnancy and to coincide with first signs of secretory activity in the acinar cells (Fig. 1). If, for instance, mammary fragments from the second part of pregnancy or early lactation are explanted in a medium containing ovarian-hypophyseal hormones which normally stimulate a high rate of cell division, the acinar cells become necrotic and die. On the contrary, when immature mammary

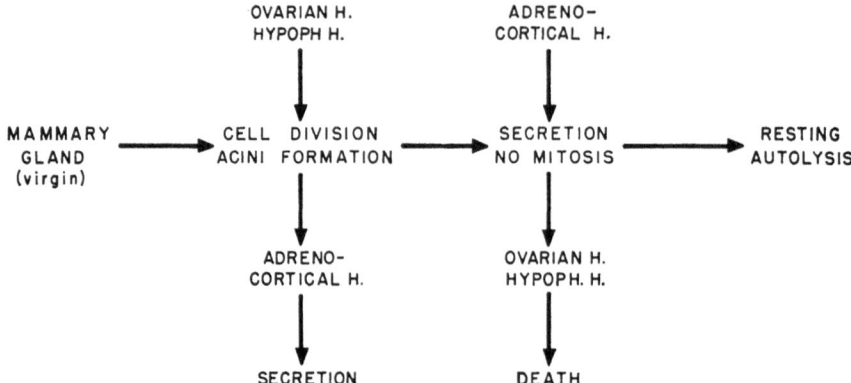

Fig. 1. Irreversibility in the differentiation pattern of the mouse mammary gland.

glands from virgin or early pregnant mice are placed in the presence of adreno-cortical hormones which currently stimulate or maintain secretion, a secretory pattern is induced. Even embryonic rudiments, when placed in these conditions, have shown a change in morphology and some ability to secrete.

It is evident, therefore, that if the immature cells can be influenced into a different pattern of behavior by environmental factors, the functional cells cannot. In fact the secretory cells appear to have reached the ultimate stage in their development with only one way out: autolysis. This is consistent with the extensive autolytic phenomena which take place in the mammary gland immediately after weaning. It is further confirmed by the relative life-span of dissociated mammary cells explanted at various positions in the pregnancy cycle.

Finite Life of the Mammary Cell

When mammary glands from agent-free mice in early lactation are dissociated with collagenases and explanted as a free cell suspension, epithelial cells cover the floor of the culture vessel in 48 hours. They originate from the acini and their cytoplasm, rich in vacuoles and fat droplets with evidence of secretory activity. This activity however is temporary; the cells do not divide and by the end of the first week of cultivation several of them begin to autolyze. Any attempt to transfer such cultures was unsuccessful (Fig. 2).

Mammary glands from the end of pregnancy behave much the same, except that autolysis sets in a little later; at the end of the second week.

In early pregnancy the majority of the cells derived from newly formed acini divide very actively. Other cell types such as fibroblasts, histiocytes, adipose cells are also found and their rapid rate of replication necessitates transfers. The maximum maintenance period for these cultures has been 7 weeks but, invariably, degenerative changes took place beyond that time.

Still less differentiated mammary cells isolated from 3-month-old virgin mice had the longest life span which averaged between 12 and 16 weeks. However, all efforts to establish a permanent mammary cell line from normal tissues failed. It

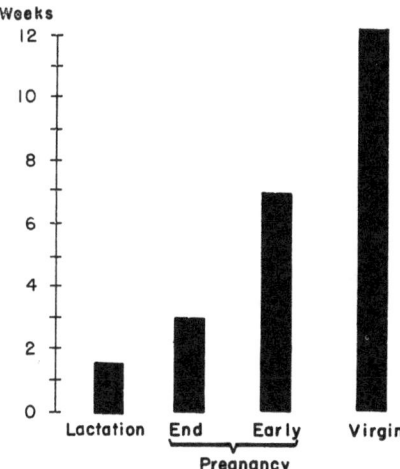

Fig. 2. Relative life span of mouse mammary tissues *in vitro*.

became evident from these results that the life of the mammary cell which, in itself is closely dependent on differentiation, is also subject to a natural finite life span which prevents continuous growth. The low growth potential of the secretory cells confirmed the irreversible character of their differentiation. On the other hand the possible finite life of the normal immature cell is much comparable to the observations of Hayflick on the finite life of human diploid cell lines (4).

Action of the Milk Agent

When primary cell suspensions from virgin agent-free mouse mammary gland are explanted, the sequence of events is always the same (Fig. 3). The epithelial cells, few in relation to the adipose and connective tissue cells, form small circular patches inter-connected with fibroblastic strands. The floor of the culture vessel is usually covered in about 4 days, after which transfers become necessary to avoid necrosis. An average of 4 transfers, each one at a week interval, can be performed

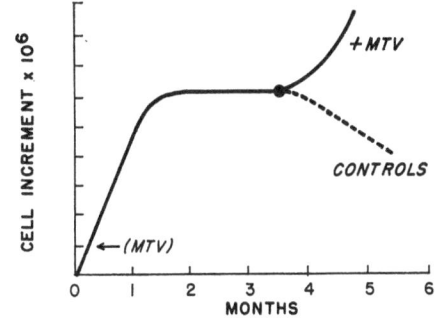

Fig. 3. Effect of MTV on the relative life span of mouse mammary tissues *in vitro*.

before the cells enter a period of complete inactivity. This period lasts about 2 months, then foci of necrosis occur spontaneously in individual cells which disintegrate leaving a granular material of cytoplasmic and nuclear origin. The number of viable cells becomes less and less from that moment until complete extinction of the culture.

The brief exposure of freshly explanted cells to the mammary tumor virus (MTV) does not sensibly change this general picture, at least until the middle or the end of the dormant period. The virus, under the form of decaseinated, defatted milk from a high cancer strain of mice is applied according to the classical procedures of virology on rapidly growing cells of 4-day-old cultures. At the end of the dormant period a few cells degenerate, while some others divide. Gradually the balance between degenerating and dividing cells changes in favor of the latter. Split transfers have to be made at an increasing rate and continuous growth is achieved.

This change in the behavior of the cultures occurs around the third month following exposure to the MTV. It is characterized, besides the continuous growth, by the appearance of giant syncytial cells and a clumping particularly noticeable in the few hours following transfer. Upon staining one can observe a frequent fragmentation of the chromatin and largely indented nuclei. Mammary cells from strain Ax mice thus transformed following exposure to milk from strain A are presently in their third year of cultivation.

Mouse embryo cells. The mammary tumor virus seems to be equally effective in transforming mouse embryo cells. The cells obtained by enzymatic dissociation of mouse embryos in the middle of pregnancy also presented a dormant period on their second and third month of cultivation. Following exposure to milk from strain RIII however, embryonic cells from C57Bl mice never entered a true inactive phase. Mitosis, although rare, did not completely disappear and on the third month, clumps of cells with a very refringent cytoplasm developed. Rapid continuous growth followed; giant multinucleated cells and nuclear deformities which had been particularly noticed in the Ax mammary cell were observed again, suggesting a similar change.

The control cultures presented a dormant period and a marked regressive phase. Unlike the mammary cells, however, complete extinction did not follow. Small isolated colonies began to expand and propagate on the 5th month of cultivation. Continuous growth was slowly established but the cultures remained completely different from the experimental series. Their rate of growth was much slower; the cells were flat, granular with a tendency to necrosis. The mitotic index was about one-fourth less than after exposure to the MTV. Although tetraploidy was observed, it was not to be compared with the numerous polyploid figures in the MTV cultures. It was, therefore, evident that the cell alteration which produced continuous growth in the control cultures was the result of an entirely different factor. In a complex system of partially differentiated cells obtained like this one, through the dissociation of whole embryos, some cells very likely have a greater capability to adapt themselves to the artificial surroundings of tissue culture. Mouse embryonic cells are prone to this kind of selectivity (16, 17) and this may be the explanation in this case. However, the difference in morphology and behavior now maintained over more than a year between the two sets of cultures is strongly indicative that a specific activity of the MTV was at the origin of the alterations observed in the experimental series.

Nature of the Transformation

The nature of the cell transformation was evaluated by implantation of cell suspensions into homologous mice. Up to 5×10^6 cells from Ax mammary cultures were inoculated subcutaneously into 3-month-old early pregnant Ax mice. Early pregnant mice presented the advantage of a natural hormonal stimulation which might be necessary to initiate a rapid cell replication. Three successive tests made 1, 2 and 6 months after exposure of the cells to the MTV have been negative. On the seventh month a fourth attempt to implant 7×10^5 cells into the glandular fat-pad itself also remained negative (Table II). The subcutaneous implants were not perceptible to palpation in the next 48 hours or at any time over a year of observation. The cells inoculated 6 and 7 months post-exposure to the agent had already shown signs of transformation *in vitro*; however, they became resorbed and could not be found at autopsy on the site of inoculation. Considering that mammary carcinoma cells from strain A mice, when inoculated in suspension into agent-free Ax animals develop as a rapidly invading tumor in 10 days, there is some doubt that the transformed culture cells had acquired a true neoplastic nature.

Similar tests made with MTV-exposed cells from C57Bl embryos and their controls have been more revealing (Table III). A subcutaneous inoculation of 1×10^6 cells a month after their contact with strain RIII milk did not have any effect on 3-month-old C57Bl mice even after 6 months of observation. On

TABLE II

AX MAMMARY CELL CULTURES EXPOSED TO MTV (A MILK) AND
IMPLANTED AS CELL SUSPENSIONS INTO ISOLOGOUS
AX MICE

per Mouse	Time Post-Exposure (months)	Assay †	Number of Mice	Time of Observ. (months)	Tumors
5×10^6	1	sc	2	12	0
5×10^6	2	sc	2	12	0
5×10^6	6 *	sc	2	12	0
7×10^5	7 *	ig	8	12	0

* Six and seven months post-exposure to the mammary tumor virus the cells were altered. No signs of alteration had been observed in the two previous tests.

† Subcutaneous (sc), intraglandular (ig).

TABLE III

C57BL MOUSE EMBRYO CELL CULTURES EXPOSED TO MTV (RIII MILK)
AND IMPLANTED AS CELL SUSPENSIONS INTO ISOLOGOUS C57BL MICE

No. of Cells per Mouse	Time Post-Exposure (months)	Assay †	Number of Mice	Time of Observ. (months)	Tumors
1×10^6	1	sc	4	2	0
1×10^6	5 *	ip	20	6	0
5×10^6	8	sc	4	5	1
5×10^6	11	sc	5	3	4

* At 5 months post-exposure the cells were altered. Control cells (not exposed to MTV) implanted in the same conditions as the exposed cells yielded negative tests in all series.

† Subcutaneous (sc), intraperitoneal (ip).

the 8th month of cultivation the same challenge produced a rapidly growing sarcoma in one of 4 inoculated mice; it developed 2 months after implantation. Another attempt was made after 11 months of cultivation, this time with 6×10^6 cells per mouse. Again, very rapidly growing sarcomas developed 2 months after this inoculation in 4 out of 5 inoculated mice; one of them had 2 tumors. Control cells, not exposed to MTV and inoculated in the very same conditions did not have any effect in the same lapse of time.

These results strongly suggest that the mouse embryonic cells which have been exposed to the MTV have undergone a true neoplastic transformation. The mammary cells, even after the morphological changes observed *in vitro*, do not appear to be cancerous. Cell alteration, therefore, does not necessarily equate with neoplasia but the data suggest that total transformation may establish itself through a gradual process. These very preliminary experiments demand further investigation and confirmation; in their light, the case of the altered Ax mammary cells should be reappraised. However, the possibility for the MTV to initiate primarily a preneoplastic state in the mammary cell is in accord with the experimental induction of preneoplastic nodules as described by DeOme (2) and Nandi (13, 14). The formation of these nodules has been found specific enough to be used in the development of quantitative assays.

Discussion and Conclusions

Whether or not the presence of the virus is necessary to insure the full neoplastic transformation of the mammary cell is not known. As a simplified working hypothesis we would like to propose the following scheme for mammary carcinogenesis in the mouse (Fig. 4).

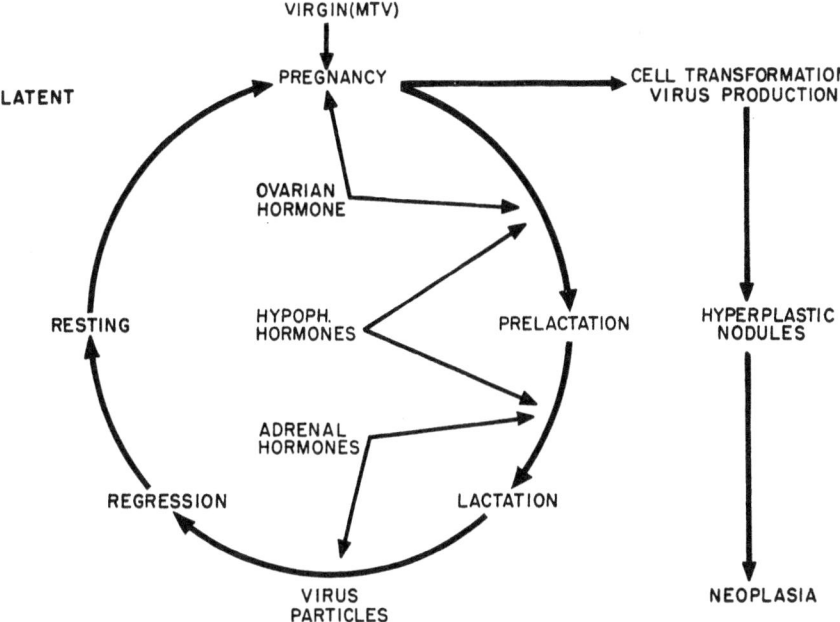

Fig. 4. The milk agent in the mammary gland cycle.

Between birth and the beginning of the first pregnancy, the mammary cell is part of an elementary duct system with a much reduced metabolism; it does not divide and has no functional activity. Yet, it is in this state that the mammary tumor virus ingested through suckling presumably enters the cell. As soon as pregnancy begins an active cell division takes place initiated by the combined stimulation of the ovarian and anterior pituitary hormones. Two alternatives are then possible depending on whether the cell harbors the virus or not: a) the mammary cell does not contain the virus—it then proceeds to the next step in differentiation which is prelactation; later, under the full impact of the adreno-cortical hormones it becomes a mature secretory cell and as such, a short-lived unit. From electron-microscope evidence, this cell can eventually pick up the virus which is then manufactured and released in large quantities at the cell-membrane level. However, the status of the secretory cell is not changed by the presence of the virus; functional differentiation being irreversible, the cell will automatically autolyze at weaning time as the gland returns to a state of rest. b) the mammary cell already contains the virus—in this stage of rapid division it becomes altered and further differentiation is prevented. This is a subtle change which does not require a sudden capability for invasion or a complete autonomy; the cell only escapes the control of differentiation but remains subject to factors governing morphogenesis (9). In organ culture these cells show an increased rate of mitosis and the walls of the acini are 2 or 3 layers thick. The number of the acini also increases significantly forming tightly packed groups which, in the animal, have been described as hyperplastic nodules. When stimulation ceases at the end of

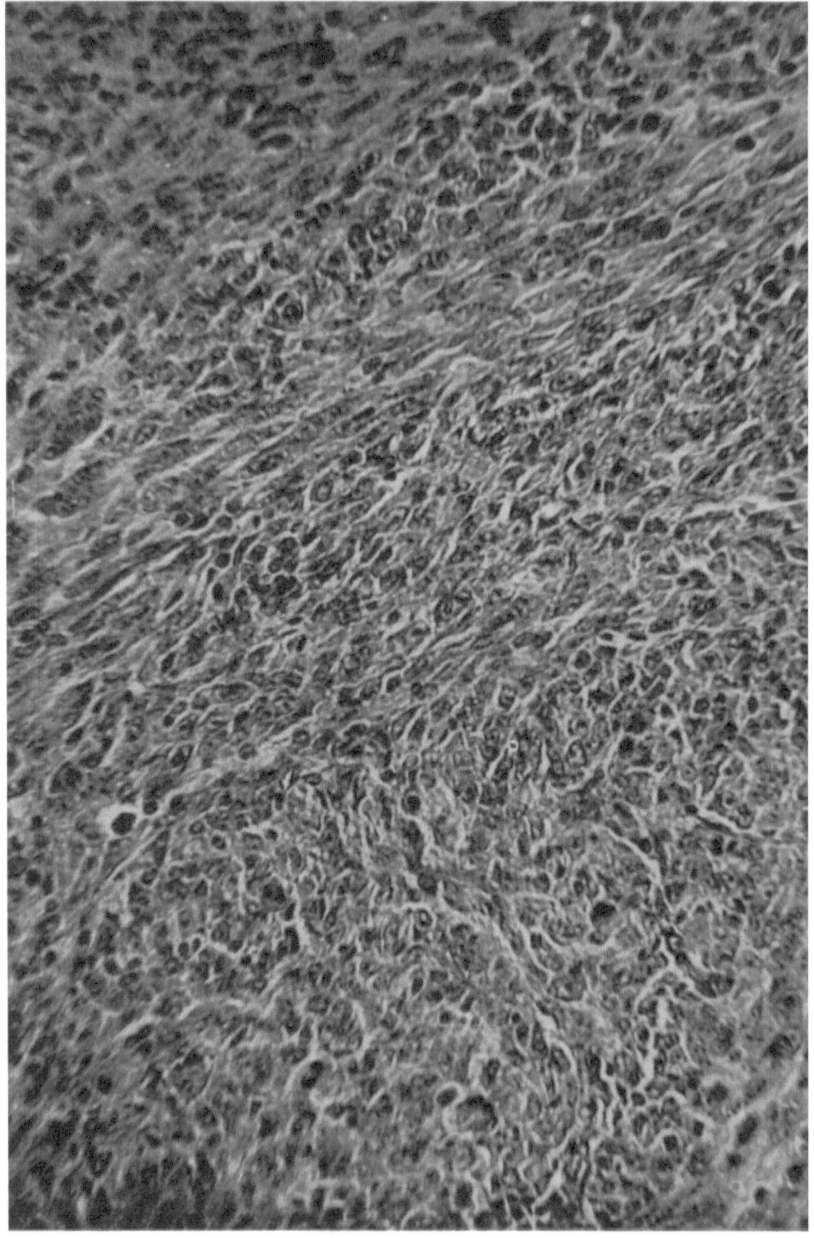

Fig. 5. Histological section of a tumor induced 2½ months after implantation of transformed C57Bl embryo cell cultures into isologous mouse. ×400.

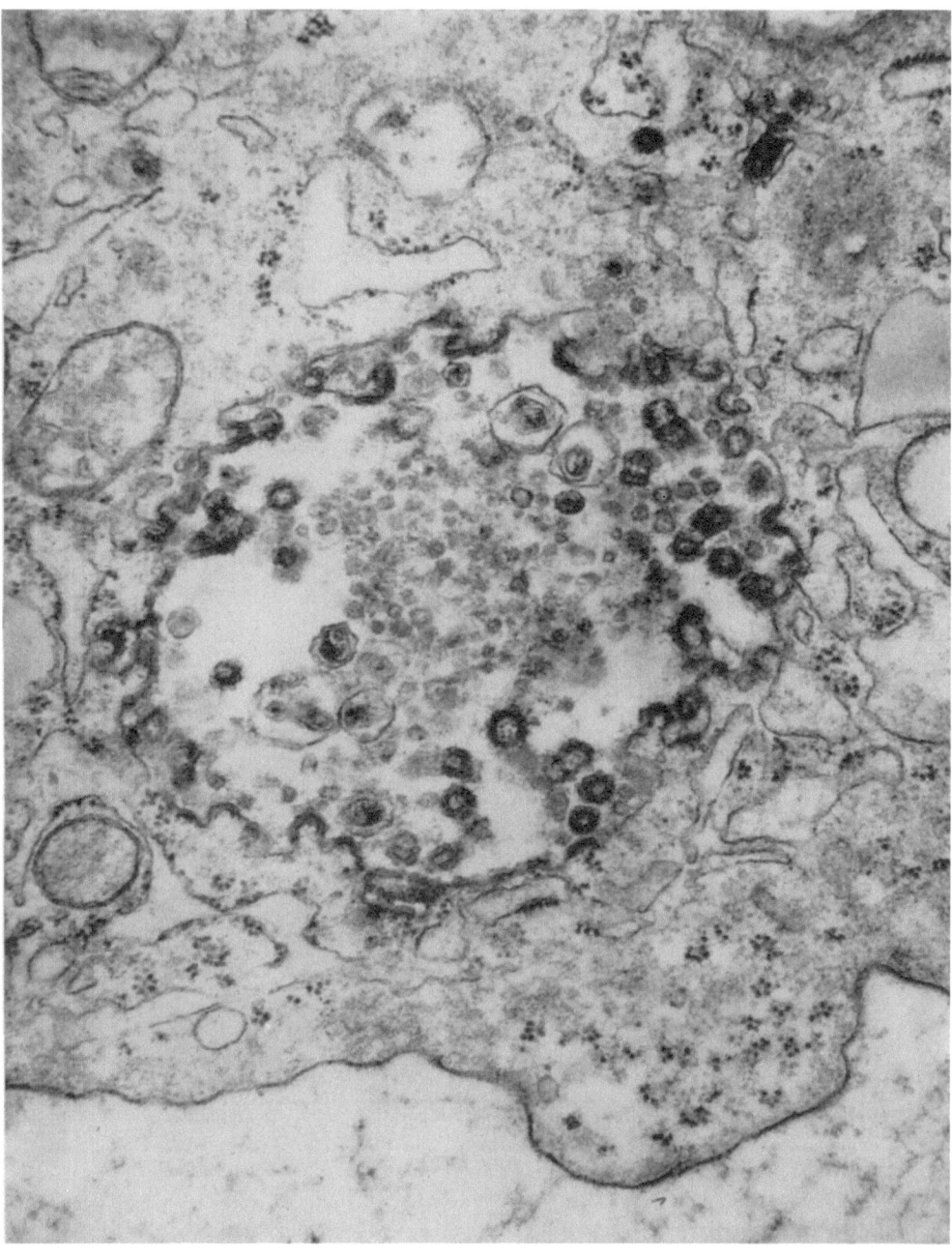

Fig. 6. Electron micrograph of the same tumor. Note the active budding of particles. ×40,000.

pregnancy or lactation, these hyperplastic acini do not autolyze like the normal ones but remain in a dormant state until stimulated again in the next pregnancy.

Complete autonomy and definite malignancy of the mammary cell probably occurs after repeated stimulation. The specific role of the virus in this final transformation is not clear. Electron-microscope studies of these tumors and of the cultures after 14 months *in vitro* have shown the presence of viral particles (Figs. 5, 6). Whether or not these particles have a tumorigenic activity is not known. An explanation for established neoplasia remains therefore a matter of speculation. It is likely, however, that given the repeated stimulations necessary to achieve this goal the life-span of the mouse is an important factor; another one could be the specific genetic sensitivity of each strain of mice.

This scheme, speculative only in its latest part, has the great advantage to give a satisfactory reason for the late appearance of mammary tumors in mice. It also points to the cell of early pregnancy as the most susceptible candidate to neoplasia. As such, the proposed mechanism may open practical possibilities for an eventual control of the disease.

References

1. COWIE, A. T.: The hormonal control of milk secretion. In Milk: The Mammary Gland and Its Secretion. S. K. Kon and A. T. Cowie, eds. Academic Press, New York, vol. 1, 168–203 (1961).
2. DeOME, K. B., FAULKIN, L. J., BERN, H. A., and BLAIR, P. B.: Development of mammary tumors from hyperplastic alveolar nodules transplanted into gland-free mammary fat pads of female C3H mouse. Cancer Res. 19, 515–520 (1959).
3. ELIAS, J. J.: Cultivation of adult mouse mammary gland in hormone enriched synthetic medium. Science 126, 842–844 (1957).
4. HAYFLICK, L.: The limited *in vitro* lifetime of human diploid cell strains. Exp. Cell Res. 37, 614–636 (1965).
5. JACOBSOHN, D.: Hormonal regulation of mammary gland growth. In Milk: The Mammary Gland and Its Secretion. S. K. Kon and A. T. Cowie, eds. Academic Press, New York, vol. 1, 127–160 (1961).
6. LASFARGUES, E. Y., and MURRAY, M. R.: Hormonal influences on the differentiation and growth of embryonic mouse mammary glands in organ cultures. Devel. Biol. 1, 413–435 (1959).
7. LASFARGUES, E. Y., and FELDMAN, D. G.: Hormonal and physiological backgrond in the production of B particles by the mouse mammary epithelium in organ cultures. Cancer Res. 23, 191–196 (1963).
8. LASFARGUES, E. Y., and MURRAY, M. R.: Comparative hormonal responses *in vitro* of mouse mammary glands from agent-carrying and agent-free strains. Acta Union Int. contre le Cancer 20, 1458–1462 (1964).
9. LASFARGUES, E. Y., MURRAY, M. R., and MOORE, D. H.: Induced epithelial hyperplasia in organ cultures of mouse mammary tissues. Effects of the milk agent. J. Nat. Cancer Inst. 34, 141–152 (1965).
10. LYONS, W. R.: Hormonal synergism in mammary growth. Proc. Roy. Soc. B. 144, 303–325 (1958).
11. MEITES, J.: Farm animals: hormonal induction of lactation. In Milk: The Mammary Gland and Its Secretion. S. K. Kon and A. T. Cowie, eds. Academic Press, New York, vol. 1, 321–367 (1961).

12. NANDI, S.: Endocrine control of mammary gland development and function in the C3H/Hc Crgl mouse. J. Nat. Cancer Inst. 21, 1039–1063 (1958).
13. NANDI, S.: New method for detection of mouse mammary tumor virus. I. Influence of foster-nursing on incidence of hyperplastic mammary nodules in BALB/c Crgl mice. J. Nat. Cancer Inst. 31, 57–79 (1963).
14. NANDI, S.: New method for detection of mouse mammary tumor virus. II. Effect of administration of lactating mammary tissue extracts on incidence of hyperplastic mammary nodules in BALB/c Crgl mice. J. Nat. Cancer Inst. 31, 75–89 (1963).
15. PROP, F. J. A.: Development of alveoli in organ cultures of total mammary glands of six weeks' old virgin mice. Exp. Cell Res. 20, 256–258 (1960).
16. ROTHFELS, K. H., KUPELWIESER, E. B., and PARKER, R. C.: Effects of x-irradiated feeder layers on mitotic activity and development of anenploidy in mouse embryo cells *in vitro*. In: Canadian Cancer Conference, vol. 5, Begg, R. W. (ed.) Academic Press, New York, 191–223 (1963).
17. TODARO, G. J., and GREEN, H.: Quantitative studies of the growth of mouse embryo cells in culture and their development into established cell lines. J. Cell Biol. 17, 299–315 (1963).

Discussion *

By K. B. DeOme

Department of Zoology and the Cancer Research Genetics Laboratory,
University of California, Berkeley, California

The paper presented by Lasfargues and Moore (11) raises many interesting possibilities. The possibility of achieving transformation in cell cultures specific for the mammary tumor virus (MTV), and of developing a quantitative assay system *in vitro*, especially interests those working with the mouse mammary tumor system.

The morphologic changes reported (11) in mammary gland cell cultures derived from Ax mice and exposed to MTV, were thought to represent transformations. Samples of these transformed cultures, however, did not give rise to mammary neoplasm following transplantation into isologous mice.

If the transformations cited (11) are analogous to the changes associated with the neoplastic transformation *in vivo* in the mouse mammary tumor system, then at least two different kinds of transformations would be expected. The first kind of transformation would represent the change from normal mammary gland cells to the cells comprising the pre-neoplastic, hyperplastic alveolar nodules. The second type of transformation would represent the change from nodule cells to neoplastic cells (8). The sequence of changes seen in the mouse mammary tumor systems can be represented in the following schema.

* Supported by USPHS grants CA-05388, CA-07015, and CA-5045, and by Cancer Research Funds of the University of California.

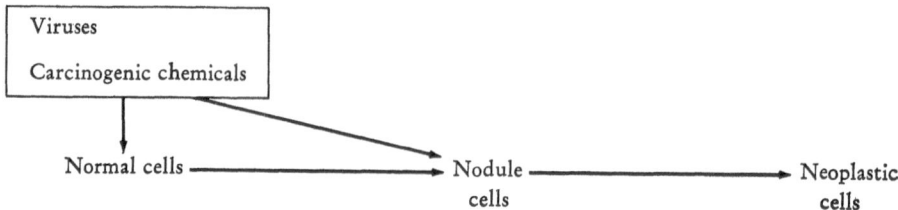

Available evidence indicates that neoplastic cells arise from nodule cells rather than from normal cells. Recent evidence shows that MTV and chemical carcinogens are active both in the change from normal cells to nodule cells and in the change from nodule cells to neo-plastic cells (12).

The transformed mammary gland cell cultures described by Lasfargues and Moore (11) apparently did not represent the change from nodule cells to neoplastic cells since samples transplanted either subcutaneously or into the "glandular fat-pads" failed to produce mam-mary gland neoplasms. It is possible, however, that these transformed cultures did represent the change from normal cells to nodular cells. Had samples of the transformed culture been transplanted into the gland-free fat-pads of young isologous female mice, outgrowth re-vealing the nature of the cultures would have been expected (9). Normal cells should have given rise to normal outgrowth, whereas nodule cells (transformed cells) should have given rise to nodule outgrowths. Unfortunately, evidence of this kind was not reported.

Abnormal outgrowths, however, have been observed following the transplantation into gland-free fat-pads of cultured cells derived from MTV-C57BL and BALB/c mammary tissues (6). In addition, cell cultures derived from two nodule outgrowths did not show unusual morphologic features.

Lasfargues and Moore report (11) a finite life span for normal mammary gland cell cultures not exposed to the MTV, and the establishment of transformed, continuous cell cultures following exposure to the MTV. A comparison of these findings with the behavior of normal and nodule cell populations *in vivo* may be instructive.

The serial transplantation of normal mammary gland tissues into the gland-free fat-pads of young isologous female mice clearly demonstrates that normal mammary cells do have a finite life span *in vivo*. Following four to six serial transplant generations, the normal cells failed to grow and further transplantation was impossible. This experiment has been re-peated using normal mammary tissues from C3Hf/Crgl, C3H/Crgl, BALB/cCrgl and C57BL/Crgl mice. In each case normal cells failed to grow for more than six transplant generations.

The finite life span of normal mammary tissues *in vivo*, however, is not related to the presence or absence of the mammary tumor viruses. C3Hf mice contain the nodule-inducing virus (NIV) and are MTV−, C3H mice are MTV+ NIV+, BALB/c mice are MTV− NIV−, and C57BL mice are MTV− and probably are NIV− (7).

Outgrowths from hyperplastic alveolar nodules, on the other hand, grow well *in vivo* for long periods of time. Two serially transplanted nodule outgrowth lines, designated B and F respectively, were derived from a nodule taken from a C3H female and were trans-planted for the first time in April 1958 (4). These two lines currently are growing well after thirty transplant generations. Another nodule outgrowth line derived from a nodule taken from a C3Hf mouse has been transplanted eighteen times over a period of more than six years (7). A hormone-induced nodule, removed from BALB/c mammary tissue, has been transplanted serially seven times over a period of thirty months (12). The absence of a short finite life span in nodule-cell populations *in vivo* is not dependent upon a particular viral association or upon the presence or absence of the mammary tumor viruses.

The relationship between the state of differentiation of the mammary epithelial cells and the occurrence of the neoplastic transformation *in vitro* suggested by Lasfargues and Moore, can be compared with results obtained with similar tissues *in vivo*. It can be shown that normal mammary gland outgrowths develop in the cleared fat-pads of young mice following the transplantation of samples taken from any part of the donor mammary gland (7). Embryonic mammary rudiments, end buds, ducts, lobules, and bits of alveolar epithelium from lactating females, produce normal outgrowths. Furthermore, cell cultures derived from mammary tissue taken from mice during late pregnancy can be retransplanted into gland-free fat-pads, wherein normal mammary gland outgrowths develop (6). These observations show that even the most differentiated mammary gland cells are terminal only if left *in situ*. They have not, however, lost their ability to grow. When placed in cleared fat-pads under appropriate hormonal conditions they replicate the series of differentiative steps which characterize the normal mammary gland.

It seems unlikely that the normal-to-nodule transformation is associated primarily with poorly differentiated mammary cells. Normal lobules arise from ducts and in most strains of mice it is probable that many nodules also arise directly from duct cells. In MTV+ strain A females, however, nodules arise only during pregnancy and lactation (3), suggesting that nodules can arise from lobules as well as from duct cells. Furthermore, nodules arise in MTV− NIV− BALB/c f C3H mice, stimulated to lobuloalveolar development with estradiol and deoxycorticosterone acetate (14), in MTV− NIV+ C3Hf mice bearing pituitary transplants for long periods of time (13), or in MTV− NIV− BALB/c mice stimulated with estradiol and deoxycorticosterone acetate and given chemical carcinogens (10).

The neoplastic transformation occurs in nodule cell populations (9). Nodules are lobule-like structures possessing hormonal requirements which permit their development and/or survival in non-pregnant, non-lactating females (2). Nodules resemble the lobules of pregnant or lactating mice (1). In virgin hosts, nodule outgrowths exhibit varying degrees of secretory activity ranging from that characteristic of mid-pregnancy to a high degree of secretory activity (4). Furthermore, the tumor-producing capabilities of nodule outgrowths is not correlated with their degree of secretory activity (4). The neoplastic transformation of nodule-cells to tumor-cells, therefore, cannot be associated with the degree of differentiation as measured by the degree of lobuloalveolar development or by extent of secretory activity. Finally, it is not surprising that cell cultures derived from lactating mammary glands should survive for shorter periods of time than cultures derived from non-lactating glands when both are maintained in media not supplemented with appropriate hormones.

Before closing this brief discussion, a few words should be said concerning our current concept of the behavior of the mammary tumor virus (MTV). Completed B particles, visualized by the electron microscope, are budded from the surfaces of mammary gland cells into the alveolar lumens (17) and probably from the epithelial cells of certain male sex accessories as well. Nodule cells and mammary tumor cells produce abundant B particles; in addition, similar B particles may accumulate in vacuoles within mammary epithelial cells (17). In addition, mammary gland tissues from infected mice contain biologically active MTV. Biologically active MTV is carried in the circulating erythrocytes, but this viral activity is not associated with the presence of demonstrable B particles, at least in MTV+ NIV− BALB/c f. C3H mice (16). BALB/c young, infected with MTV by the ingestion of milk containing B particles, carry blood-borne activity throughout their lives. Their mammary gland tissues, however, do not show evidence of infection until the animals exceed twelve weeks of age. These observations suggest that MTV must exist in at least two forms;

one form associated with B particles and another form not associated with B particles. The formation of B particles in mammary gland tissues and in male sex accessories may be associated with the transmission of the virus from parent to offspring. The blood-borne activity not associated with B particles may represent the form of the MTV which infects various tissues within the body of the infected mouse.

This view is supported by additional observations. For example, the blood-borne MTV activity is highly strain specific, whereas the B particle-associated activity is not highly strain specific (15). Similarly, the B particle-associated MTV activity is transmissible by the oral route, whereas the blood-borne activity is not transmitted readily by the oral route (16).

The reported production of sarcomata following the transplantation into isologous C57BL mice of embryo cell cultures exposed to MTV as long as eight months prior to transplantation, represents a very interesting observation. Since mouse embryos are not normally exposed to the MTV, an *in vivo* model is not available. These "very preliminary results" require extensive confirmation. Many alternate explanations must be eliminated. For example, Sandford reported the occurrence of the neoplastic transformation in cell cultures of fibroblasts derived from C3H mice. Similarly, sarcomata were occasionally produced *in vivo* following the injection of the Gross leukemia virus and the polyoma virus. In spite of the difficulties, we can hope that an *in vitro* assay system for the MTV can be developed.

References

1. BERN, H. A., DeOME, K. B., ALFERT, M., and PITELKA, D. R.: Morphologic and physiologic characterization of hyperplastic nodules in the mammary gland of the C3H/He Crgl mouse. Proceedings of the II International Symposium on Mammary Cancer, Perugia, July 24–29, pp. 565–573 (1957).
2. BERN, HOWARD A., and NANDI, S.: Recent studies of the hormonal influence in mouse mammary tumorigenesis. Progress in Experimental Tumor Research 2, 90–144 (S. Karger, Basel/New York, 1961).
3. BLAIR, P. B.: *Unpublished data.*
4. BLAIR, PHYLLIS B., DeOME, KENNETH B., and NANDI, SATYABRATA: The preneoplastic state in mouse mammary carcinogenesis: In: Henry Ford Hospital Symposium: Biological Interactions in Normal and Neoplastic Growth, pp. 371–389, Little, Brown and Co., Boston (1962).
5. DANIEL, C. W.: *Unpublished data.*
6. DANIEL, C. W., and DeOME, K. B.: Growth of mouse mammary glands *in vivo* after monolayer culture. Science 149, 634–636 (1965).
7. DeOME, K. B.: *Unpublished data.*
8. DeOME, K. B., BERN, H. A., NANDI, S., PITELKA, D. R., and FAULKIN, L. J., JR.: The precancerous nature of the hyperplastic alveolar nodules found in the mammary glands of old female C3H/He Crgl mice. M. D. Anderson Hospital and Tumor Institute: Genetics and Cancer, University of Texas Press, Houston, pp. 327–348 (1959).
9. DeOME, K. B., FAULKIN, L. J., JR., BERN, H. A., and BLAIR, P. B.: Development of mammary tumors from hyperplastic alveolar nodules transplanted into gland-free mammary fat pads of female C3H mice. Cancer Res. 19, 515–520 (1959).
10. FAULKIN, L. J., JR.: Hyperplastic lesions of mouse mammary glands after treatment with 3-methylcholanthrene. J. Nat. Cancer Inst. 36, 289–297 (1966).
11. LASFARGUES, E. Y., and MOORE, D. H.: Cell transformation by the mammary agent in tissue culture. Symposium on the Malignant Transformation. University of Chicago School of Medicine (1966).

12. MEDINA, D., and DeOME, K. B.: *Unpublished data.*

13. MÜHLBOCK, O., and BOOT, L. M.: Induction of mammary cancer in mice without the mammary tumor agent by isografts of hypophyses. Cancer Res. 9, 402–412 (1959).

14. NANDI, S.: New method for the detection of mouse mammary tumor virus. I. Influence of foster nursing on incidence of hyperplastic mammary nodules in BALB/cCrgl mice. J. Nat. Cancer Inst. 31, 57–73 (1963).

15. NANDI, S., HANDIN, M., and YOUNG, L.: Strain specific mammary tumor virus activity in blood of C3H and BALB/c *f*. C3H strains of mice. J. Nat. Cancer Inst. (*In press*)

16. NANDI, S., KNOX, D., DeOME, K. B., HANDIN, M., FINSTER, V. V., and PICKETT, P. B.: Mammary tumor virus activity in red blood cells of BALB/c *f*. C3H mice. J. Nat. Cancer Inst. (*In press*)

17. PITELKA, D. R., DeOME, K. B., and BERN, H. A.: Viruslike particles in precancerous hyperplastic mammary tissues of C3H and C3Hf mice. J. Nat. Cancer Inst. 25, 753–777 (1960).

Specific Antigens Produced by Oncogenic Viruses

By

Karl Habel [1]

Although it has been known for many years that certain viruses have something to do with tumor induction, in most instances there has been no specific evidence that the individual tumor cell had been directly transformed by a given virus. This sort of evidence was available only in the case of those RNA viruses, such as the avian leukoses, where the infected transformed cell continues to produce infectious virus. In the DNA virus-induced tumors, which are frequently free of infectious virus, there was no easy, direct way of establishing the etiological relationship. However, with the discovery in recent years that virus-induced tumors contain specific new antigens, a ready means of identification became available. There is good reason to believe that these antigens will be present in all virus-induced tumors, since one type or another has been found in all classes of tumors thus far adequately examined.

Classification of Antigens

As summarized in Table I, the new, specific antigens demonstrable in virus-induced tumors and cells transformed *in vitro* by viruses may be divided on a

TABLE I

SPECIFIC ANTIGENS IN VIRUS-TRANSFORMED CELLS

	RNA (RSV, leukemia)	DNA (Papova, adeno)	Technics
Cellular Antigens			
Surface (transplantation)	+	+	FA, cytotoxicity, transpl. immunity
Internal (serological)	+	+	FA, CF, agar diff., cytotoxicity
Viral Antigens	+	−	FA, CF, agar diff.

[1] United States Department of Health, Education, and Welfare, Public Health Service, National Institutes of Health, National Institute of Allergy and Infectious Diseases, Laboratory of Biology of Viruses, Bethesda, Maryland.

variety of bases. Certainly viral antigens consisting of structural proteins, which are a part of the intact, mature, virion should be distinguished from those induced by the virus, but not physically incorporated in the virus particle. Since there are major differences in the virus-cell relationships in tumors induced by RNA and DNA viruses, this too is a means of comparing the antigens. Differences also exist in the tumor-host relationships as well as in the applicability of certain immunological tests between solid tumors and leukemias. However, the most practical division of the virus-induced antigens is based on the methods used to demonstrate their presence —the transplantation types versus those requiring some serological type of test.

Viral Antigens

Antigens which are structural parts of the mature virus particle may be demonstrated in certain tumor cells. These antigens usually represent coat proteins and they may occur in a free form before assembly into virions, as empty organized outer viral shells or as complete virions.

Demonstration of this type of antigen is usually limited to cells transformed by RNA viruses, such as the avian (72) and murine leukoses (9). These viruses mature at the cell surface in a manner similar to standard myxoviruses and when fluorescent antibody (FA) staining technics are used the antigen is usually located at the cell surface. However, not all leukosis cells contain these viral coat antigens since chicken fibroblasts transformed *in vitro* by Bryan strain of Rous Sarcoma Virus (RSV) in the absence of a helper virus do not produce new, mature, virus particles (29), yet they do contain demonstrable internal viral antigens (73).

Tumors induced by DNA viruses are not always free of viral antigens. A few cells in polyoma (46), and SV40 tumors (11), may support a lytic cycle of virus replication with the appearance of new infectious virus and the corresponding viral antigens. However, this is probably accompanied by cell death and lysis. Shope papillomas in rabbits induced by a DNA virus contain mature infectious virus and FA staining viral antigens only when the tumor cells become cornified and die at the surface of the tumor (47). In fact, very early work with the Shope virus showed the development of neutralizing antiviral antibodies in domestic rabbits where the tumor is free of demonstrable virus (38). An early report demonstrated the adenovirus "C" antigen in adenovirus 12 tumors of hamsters (35), but later studies have raised questions concerning this finding (13). Extensive attempts to find viral antigens in virus-free polyoma transformed cells have been completely negative (17, 71).

The viral antigens can be demonstrated by a number of serological technics but FA staining is probably the most useful method. Complement fixation (CF), hemagglutination (HA), production of neutralizing antibodies, agar diffusion and coprecipitation of radioactively-labeled antigen are all applicable in certain situations.

The significance of the viral antigens in oncogenesis is not known but their frequent localization in the cell membrane makes them potentially important since this structure is involved in such things as contact inhibition, cell division and immunological reactions. They, therefore, may be involved in transplantation resistance (18).

Transplantation Type Antigen

For many years two consistently observed phenomena strongly suggested some type of immunological reaction between the host animal and his own virus-induced tumor. The first was the eventual disappearance of many tumors after rapid growth to a certain size and the second was the strong age dependence of tumor production after virus inoculation.

The transplantation type of antigen has been demonstrated only indirectly through resistance to tumor transplant challenge and was first shown with the polyoma virus-induced tumors of mice and hamsters (18, 60). Although all types of virus-induced solid tumors and leukemias appear to contain the transplantation type of antigen as demonstrated by immunization with virus or tumor against tumor challenge, most of our knowledge of this type antigen comes from studies on the DNA papova viruses, especially polyoma. A single dose of polyoma virus inoculated into adult animals will multiply in various organs and call forth an antiviral antibody response in the serum, but will not produce symptoms or tumors. These virus-immune adults will be resistant to challenge with a "virus-free," transplantable, isologous polyoma tumor. This resistance is only relative and may be overcome by challenge with a sufficiently large number of tumor cells, but tumor development in this situation is definitely retarded. Resistance can be produced by immunization with the virus-free tumor as well as with live virus. This together with the fact that the resistance can be transferred by viable lymph node cells from a resistant isologous mouse (19, 59) suggested that a new antigen of the homotransplantation type was present in the virus-induced tumor, and that resistance to the tumor challenge was due to a homograft type of rejection. Also consistent with this explanation of experimental findings was the evidence that virus-immune animals resisting one level of tumor challenge were subsequently able to reject a higher level (19).

One of the most important characteristics of this new transplantation type of antigen produced in virus-transformed cells is its specificity. Animals made resistant to challenge with polyoma tumors are completely susceptible to challenge with non-polyoma tumors (18, 60). In experiments showing that the same resistance phenomenon can be demonstrated in hamsters inoculated with infectious SV40 virus and challenged with a transplantable SV40 hamster tumor, the specificity was readily demonstrated by cross-challenge tests (6, 23, 43). Adult hamsters inoculated with polyoma virus resisted challenge with a polyoma tumor but not with an SV40 tumor, while SV40 virus inoculated animals resisted the SV40 tumor but not the polyoma tumor. It is of interest that a similar transplantation antigen has been described in carcinogen-induced tumors, but here there is no specificity and no common antigenicity between tumors caused by the same agent (40).

The nature of the transplantation type antigen in polyoma virus-induced tumors is not known. Most attempts to produce cytotoxic antibodies against tumor cells have been negative or equivocal (2), and limited tests for immune adherence or lytic effect with tumor cells exposed to lymphocytes from resistant animals have given negative results. However, there have been recent reports of FA staining of cell membranes of virus-free SV40 tumor cells (67) and cytotoxic antibody lysis of polyoma transformed cells (32) which may very well represent a demonstration of the transplantation type of antigen. All the evidence to date indicates that the antigen is cellular and not viral in the sense of being a structural part of the infectious virus particle. Serum from animals made resistant to tumor challenge by immunization with virus-free polyoma tumors is negative for neutralizing, hem-agglutination-inhibiting (HI) and complement-fixing (CF) antiviral antibodies. Adsorption of an antiviral serum with virus-free tumor cells does not reduce the titer of antibodies for viral antigens. Although all virus-immunized animals have high titers of antiviral antibodies at the time they are resistant to tumor challenge, these antibodies do not appear to be responsible for the resistance. Newborns having circulating antiviral antibodies derived by transplacental transfer from their immune mothers are not resistant to challenge with tumor (19).

The fundamental character of the appearance of the transplantation antigen in polyoma-induced tumor cells is supported by the fact that cells transformed *in vitro* by this virus also contain the antigen (20).

Not only is the transplantation antigen specific but it appears to be an expression of a genetically stable character of the transformed cell. Polyoma mouse tumors can be shown to be positive for the antigen after many transplant passages even when selective pressure has been applied against the antigen by passing high doses of tumor cells through virus-immune mice (61). Likewise, mouse embryo cells transformed *in vitro* by polyoma continue to have the antigen after several years of cultivation (20). On the other hand, hamster polyoma tumors and hamster embryo cells transformed *in vitro* tend on passage to show a quantitative reduction in their ability to be rejected by virus-immune animals (21).

The question as to whether the transplantation antigens induced in polyoma tumors of the mouse and the hamster are the same has not been completely resolved. There is some evidence of a relationship but technical difficulties in quantitative experiments in heterotransplantation systems have made results difficult to interpret. Further evidence comes from a somewhat different technic. Girardi and associates had shown that hamsters inoculated with SV40 virus as newborns and at maturity with X-irradiated SV40 hamster tumor cells had a reduced incidence of primary tumors (16). Subsequent experiments showed the same type of protection when unirradiated human cells transformed by SV40 were used for immunization (14). Attempts to find if cells undergoing a lytic polyoma infection might also produce the transplantation antigen have not given clear-cut results.

Perhaps one of the most significant aspects of the experimentation on the virus-induced transplantation antigens has been its implication for the oncogenic properties of polyoma virus under conditions of both experimental and natural infection. My initial investigation of cellular antigens and immunological reactions in the polyoma system was prompted by a desire to explain why oncogenesis by most tumor viruses is dependent upon the age of the virus-inoculated host.

The presence of a new transplantation antigen in polyoma tumors provided a logical basis for explaining the resistance to tumor challenge developing in adult mice and hamsters after immunization with a single inoculation of infectious virus. When polyoma virus is inoculated into adults there is not only virus multiplication with lytic destruction of cells but also cell transformation followed by multiplication of the tumor cells which now contain a new antigen. The immunologically competent adult recognizes the new antigen and eventually rejects its own developing tumor in a homograft reaction. Having thus become sensitized to the virus-induced tumor antigen, the virus-immunized animal is capable of resisting challenge with a transplantable isologous polyoma tumor containing the same antigen by a "second set" homograft reaction. On the other hand, the transformation of normal cells to tumor cells containing a new transplantation type antigen in the immunologically incompetent newborn animal causes no rejection and tumor development ensues. After the demonstration of a new transplantation antigen in polyoma tumor cells, an experiment was designed to test whether the inefficient immunological response of newborn mice could be responsible for the fact that these immature animals develop tumors after polyoma virus infection, whereas adults do not (22). Large groups of newborn and adult mice were inoculated with high doses of polyoma virus and at intervals groups of animals were bled, their kidneys removed and others quantitatively challenged with a transplantable polyoma tumor. Serum was tested for antiviral HI antibody and kidneys for infectious virus. Adult animals rapidly produced new infectious virus and antiviral antibodies and within 3 days showed resistance to tumor challenge. The virus-inoculated newborns also had rapid virus multiplication and a somewhat slower antiviral antibody production but in striking contrast to the adults did not develop resistance to tumor challenge until the 26th day, indicating at least a short-term tolerance for the new transplantation tumor antigen.

The fact that newborns inoculated with virus do eventually develop resistance to tumor challenge even though they are in the latent period of their own tumor development is somewhat difficult to explain. However, it is known that a 100% incidence of tumor induction in newborn mice inoculated with large doses of polyoma virus is not consistently obtained, and when a titration of oncogenic capacity of a virus preparation is made in mice it is not unusual to have skips occurring irregularly over several logs' differences in dose. It would appear that after virus inoculation of the newborn animal there are at least two dynamic systems competing—the multiplication of transformed cells to give a critical antigenic mass too great for immunological rejection and the rapidly maturing immunological system of the animal with eventual homograft reaction to its own transformed cells. The relative balance of these two systems with time must determine the outcome in terms of the appearance of a gross tumor induced by the polyoma virus inoculated into the newborn.

The experimental finding that whole body X-irradiation (45) or thymectomy (70) will make adult animals susceptible to tumor production by polyoma virus, is strong evidence supporting the thesis for immunological control of polyoma tumor development. For years it has been known that polyoma virus infection is widespread under natural conditions of exposure in both laboratory mouse colonies (54) and in wild mice (55), yet a naturally occurring polyoma tumor is an extreme rarity. A logical explanation is that mice become exposed at a time when they are immunologically competent or that they are exposed to such small doses of virus as newborns that the immunological balance is predominantly in favor of tumor rejection. A very interesting confirmation of this explanation has recently been reported by Law (44), who found that mice thymectomized as newborns and kept in a room in which polyoma infection was spreading by exposure under natural conditions, did indeed develop polyoma tumors.

As mentioned before, a variety of experiments have indicated a transplantation type of new antigen in every virus-induced tumor in which it has been sought. The development of resistance to tumor challenge following inoculation of virus into adults as used in the polyoma system has since been demonstrated in SV40 (23, 6, 43), adenovirus 12 (69) and Moloney leukemia (58). Resistance to isograft challenge following immunization with tumor allografts, extracts or X-irradiated cells has been used to demonstrate the transplantation antigen in Schmidt-Ruppin RSV virus-induced mouse tumors (37), Shope papillomas in rabbits (8), Bittner virus mammary tumors (75), and in Gross (1, 42, 74), Moloney (58, 48, 15, 41), Rauscher (48, 4) and Graffi (50) virus leukemias.

Serological Type Antigens

The use of serum antibodies to demonstrate specific new non-viral antigens in virus-induced tumors was initiated by Zilber and his associates with antisera produced in heterologous species against Rous sarcomas (76). Most recent studies have used homologous antisera in the FA staining technic, the cytotoxic test or by CF. The most universally applicable methods are the FA and CF tests, since the cytotoxic reaction in general is limited to leukemia or lymphoma cells. All three procedures have now been used to study tumors induced by RNA and DNA viruses.

RNA virus-induced tumors. Since RNA tumor viruses continue to replicate in the cells they have transformed, it is difficult to distinguish between viral antigens and newly induced cellular antigens as being responsible for serological reactions. Even in the case of murine leukemias and Bittner virus mammary tumors where transplantable tumors can be established in isologous strains, these transplants continue to elaborate intact infectious virus particles. Although this problem of determining the source of the antigen makes interpretation of results somewhat difficult these serological reactions have been most helpful in showing interrelationships and immunological factors important in oncogenesis.

Using the cytotoxic test, Old and his associates (for review see (49)) have established antigenic crossing between leukemias caused by several of the known murine leukemia viruses and have demonstrated the existence of five different antigens. The "G" antigen, characteristic of Gross leukemia, has been found in the leukemic cells and normal lymphoid tissues of 5 mouse strains having a high incidence of spontaneous leukemia. An antigen designated the "FMR" antigen has been found in common between Friend, Moloney, and Rauscher virus induced leukemias. An "ML" antigen is common between certain leukemias and mammary tumors. "E" antigen is present in certain leukemias of C57BL mice and "TL" antigen is not only found in certain leukemias but in the normal thymus of certain mouse lines. Furthermore, a non-cell-associated "soluble" specific antigen can be demonstrated in the plasma of certain murine leukemias (64) and in human leukemia (10).

The FA test has been used by the Kleins and their associates for studies of the Gross (63) and Moloney (41) leukemias and in general their findings are similar to those shown by the cytotoxic test. Even greater cross-reactivity in murine leukemias (31) and in the avian leukoses (34) have been shown by the CF test.

DNA tumor viruses. Most of the work done on serological antigens of tumors produced by these viruses has involved the FA and CF tests, although in certain situations the cytotoxic test has also given positive results. Huebner and his associates (33) were the first to demonstrate a CF antigen in adenovirus-induced hamster tumors, using antisera from hamsters carrying the tumors. The CF antigen has now been shown in several adenovirus tumors and those induced by SV40 (5) and polyoma (24) viruses. These antigens are not structural antigens of the virus particles, but are specific for tumors induced by a given virus. They are also present in cells transformed *in vitro* by these same viruses. These antigens occur in various histological types of tumors and in tumors of different species caused by the same virus. Even human cells transformed *in vitro* by SV40 virus contain relatively high titers of this type of antigen (25). Indirect evidence suggests that this CF antigen is different from the homotransplantation antigen in DNA virus tumors. In general, antibodies to this antigen are present only in animals carrying tumors and when the tumor is surgically removed the antibody levels drop. As in the different types of avian and murine leukemias, there are also certain cross reactions between antigens produced by different types of adenoviruses (36).

An interesting finding concerning the CF antigens has been their demonstration in cells undergoing a cytolytic infection with the DNA tumor viruses (13, 26, 33, 52, 57). In this situation, the antigens appear early in the virus multiplication cycle before new infectious virus or viral coat antigens and tend to decrease late in infection. The time relationships to the infectious cycle are the same for the appearance of these antigens, the increase in certain enzymes such as thymidine kinase required for DNA synthesis (39), and virus-induced cellular DNA synthesis (7). Because of this, it is tempting to postulate that the antigens may be early virus-induced enzymes involved in DNA synthesis, but direct evidence for this is not yet available.

In practically every situation where the CF antigen is found it can also be demonstrated by FA staining. Staining is most prominent in the nucleus, both in "virus-free" tumors or *in vitro* transformed cells and in cells undergoing a lytic infection (51). In the polyoma system however, FA staining of nuclei is positive in the lytic infection, but not in tumor cells (66). One of the obvious advantages of the FA technic in addition to its localizing ability, is its direct demonstration of which and how many cells are producing the antigen.

Relatively little work has been done to characterize this serological antigen by chemical, physical or immunological evidence. Berman and Rowe (3) using agar immunodiffusion technics had evidence that adenovirus tumors contain the CF antigen, viral "C" antigen and a third new type of antigen. The CF antigen is not sedimented by high speed centrifugation, appears to be heat labile, and is not destroyed by treatment with DNAase and RNAase, but is by trypsin digestion (12).

One additional serological antigen produced by a DNA virus tumor is of especial interest. Rogers (53) has shown that Shope papillomas elaborate a new species of arginase which is antigenic and calls forth specific antibodies in the tumor-bearing rabbit demonstrable by the precipitin test.

Problems in Studies on Virus-induced Tumor Antigens

Relation of antigen to original tumor-inducing virus. One of the most significant applications of the specific virus-induced tumor antigens concerns their specificity and use in identifying a given tumor as having been induced by a given virus. Such identification, however, makes at least three basic assumptions: 1) that all tumors induced by a given virus will contain the new antigen; 2) that the antigen will persist in all daughter cells, and 3) that the antigen will not be produced in cells of tumors arising from other causes when they might be superinfected by a tumor virus. Exceptions have already been found to all three of these assumptions. Although most tumors induced by polyoma virus show the homotransplantation type of antigen, some strains of virus lack the ability to establish resistance to tumor challenge (30). Likewise, in the case of the CF antigen some tumors induced by certain polyoma virus strains do not have sufficient antigen to react in the CF test against standard antisera (24). Once either type of antigen is present in a polyoma tumor there is no evidence that it disappears during the life of a particular tumor-bearing animal, but on long transplant or tissue culture passage polyoma transformed hamster cells appear to have less evidence of the transplantation antigen and may lose the CF antigen.

There has been one report that a methyl-cholanthrene (M-C) transplantable tumor purposefully contaminated with polyoma virus was resisted by polyoma-immune mice (62). Furthermore, there has been reported an even more interesting phenomenon where a tumor contaminated with herpes virus was resisted on transplantation into a herpes-immune animal (28). *In vitro* "super-transformation" of an already polyoma-transformed mouse cell on exposure to SV40 virus has given rise to clones which contain both polyoma and SV40 CF tumor antigens (68). Likewise, Stück, et al. (65) showed that a transplantable Gross leukemia could acquire another specific antigen on *in vivo* exposure to Rauscher leukemia virus.

Tumor-specific antigens produced in lytic infections. Although there is as yet no evidence concerning the transplantation-type antigen, it is now well established that cells undergoing a lytic infection with SV40, adenoviruses and polyoma contain an antigen reacting in the CF and FA staining tests with hamster antisera positive against tumors induced by these viruses. Kidneys of newborn mice removed 5 to 7 days after inoculation with polyoma virus, at which time virus replication is at its height, show the CF tumor antigen. Even adult mice inoculated with polyoma virus will occasionally develop CF antibodies which later disappear. Thus, the presence of antibodies against the tumor antigen might reflect only a reaction to antigen produced by lytic infection instead of being due to any tumor which might also be present. Although this obviously raises problems in the interpretation of positive CF tests, it also theoretically may broaden the application of serological tests in the search for possible oncogenic properties of known "ordinary" viruses. Up until now, the only CF antibodies available in testing for the presence of virus tumor antigens were in sera of animals carrying a tumor induced by a given virus. If such antibodies could be produced by immunizing animals with extracts of lytically infected cells, then tumors of unknown etiology, i.e., human tumors, might be tested for possible etiological relationship with common viruses such as herpes, measles, etc. Likewise antibodies in tumor patients might be demonstrated against antigens appearing in lytic infections with such "ordinary" viruses.

Presence of normal isoantigens. Since most of the work on the CF antigens of virus-induced tumors has been done in the non-inbred hamster, it is not surprising that a "normal" antigen has been reported as confusing the experimental findings. Sabin (56) has found that certain hamsters (and their virus-induced tumors) have a normal antigen which may be lacking in other hamsters. The latter animals will respond with CF antibodies to this antigen as well as to the tumor specific one when inoculated with a positive transplantable virus tumor. This leads to a necessity for checking all standard antigens and antisera for this complicating factor before interpreting CF and fluorescent antibody staining results (27).

Antigen-antibody interactions between species. As soon as the virus tumor specific antigens and antibodies from one species, such as the hamster, are used to test materials from other species, such as man, there is the possible difficulty of interactions between normal constituents of tissues and sera. This has already been found in the CF test (25). In tests in collaboration with a group at the Wistar Institute we found that certain cancer patients who had rejected transplants of human cells transformed by SV40 virus had high CF titers in their sera against hamster SV40

TABLE II

PROPERTIES OF TWO TYPES OF ANTIGENS PRODUCED BY DNA TUMOR VIRUSES

	Homotransplantation Antigen	CF Antigen
Present in tumor cells	+	+
Present in *in vitro* transformed cells	+	+
Present in lytic infection	?	+
Location in cell	Surface	Nucleus
Specificity	+	+
Cross reacting between species	+	+
Animal reaction—cell mediated	+	−
—serum antibody	−	+
—produced by virus inoc.	+	±
—produced by tumor inoc.	+	+
Involved in viral oncogenesis	+	?

tumor antigen. It was readily shown that pre-transplant sera were equally high titering, that the sera were also positive against other hamster tumors and normal hamster tissues but negative if human SV40 transformed cells were used as antigen.

Discussion

That new, specific antigens appear in cells transformed by tumor viruses is now well established and this is probably a general phenomenon. Although direct evidence is lacking, it is logical to assume that their production is coded by a persisting viral genome or a segment thereof. The outstanding feature of these antigens is their specificity which is determined by the original inducing virus and their presence in all types of tumors and cell lines, even in widely divergent species.

In spite of certain pitfalls and problems in their application, these antigens have been and will continue to be found very useful in establishing an etiological relationship between a given virus and a tumor induced by it. At the least, the demonstration of such antigens signifies a dynamic interaction between the virus and the cell in which the antigen is found.

The significance of these antigens in the mechanism of virus oncogenesis has yet to be delineated. However, it is already obvious that the transplantation antigen and the animals' immunological reaction to it are strong determinants of gross tumor development, eventually resulting from transformation of cells by tumor viruses. The role of serological antigens is not so obvious. Some may have no role in oncogenesis but it is interesting to speculate that they have some function in the replication of viral nucleic acids, or even perhaps in repression of normal mechanisms controlling cell DNA synthesis and cell multiplication.

References

1. AXELRAD, A. A.: Changes in resistance to the proliferation of isotransplanted Gross virus-induced lymphoma cells, as measured with a spleen colony assay. Nature 199, 80 (1963).
2. BASSES, R.: Antigenic differences between normal and polyoma virus-transformed cells. II. In vitro evidence for a virus-induced antigen. Cancer Res. 24, 1216 (1964).
3. BERMAN, L. D., and ROWE, W. P.: A study of the antigens involved in adenovirus 12 tumorigenesis by immunodiffusion techniques. J. Exptl. Med. 121, 955 (1965).
4. BIANCO, A. R., GLYNN, J. P., and GOLDIN, A.: Immunity against isotransplants of Rauscher virus-induced leukemia. Proc. Am. Assoc. Cancer Res. 5, 5 (1965).
5. BLACK, P. H., ROWE, W. P., TURNER, H. C., and HUEBNER, R. J.: A specific complement-fixing antigen present in SV40 tumor and transformed cells. Proc. Natl. Acad. Sci. 50, 1148 (1963).
6. DEFENDI, V.: Effect of SV40 virus immunization of growth of transplantable SV40 and polyoma virus tumors in hamsters. Proc. Soc. Exptl. Biol. Med. 113, 12 (1963).
7. DULBECCO, R., HARTWELL, L. H., and VOGT, M.: Induction of cellular DNA synthesis by polyoma virus. Proc. Natl. Acad. Sci. 53, 403 (1965).
8. EVANS, C. A., WEISER, R. S., and ITO, Y.: Antiviral and antitumor immunologic mechanisms operative in the Shope papilloma-carcinoma system. Cold Spring Harbor Symposium on Quant. Biol. 27, 453 (1962).

9. FINK, M. A., and MALMGREN, R. A.: Fluorescent antibody studies of the viral antigen in a murine leukemia (Rauscher). J. Natl. Cancer Inst. **31**, 1111 (1963).

10. FINK, M. A., MALMGREN, R. A., RAUSCHER, F. J., ORR, H. C., and KARON, M.: Application of immunofluorescence to the study of human leukemia. J. Natl. Cancer Inst. **33**, 581 (1964).

11. GERBER, P.: Virogenic hamster tumor cells: Induction of virus synthesis. Science **145**, 833 (1964).

12. GILDEN, R. V., CARP, R. I., TAGUCHI, F., and DEFENDI, V.: The nature and localization of the SV40-induced complement-fixing antigen. Proc. Natl. Acad. Sci. **53**, 684 (1965).

13. GILEAD, Z., and GINSBERG, H. S.: Characterization of a tumorlike antigen in type 12 and type 18 adenovirus-infected cells. J. Bacteriol. **90**, 120 (1965).

14. GIRARDI, A. J.: Prevention of SV40 virus oncogenesis in hamsters. I. Tumor resistance induced by human cells transformed by SV40. Proc. Natl. Acad. Sci. **54**, 445 (1965).

15. GLYNN, J. P., BIANCO, A. R., and GOLDIN, A.: Studies on induced resistance against isotransplants of virus-induced leukemia. Cancer Research **24**, 502 (1964).

16. GOLDNER, H., GIRARDI, A. J., LARSON, V. M., and HILLEMAN, M. R.: Interruption of SV40 virus tumorgenesis using irradiated homologous tumor antigen. Proc. Soc. Exptl. Biol. Med. **117**, 851 (1964).

17. HABEL, K., and SILVERBERG, R. J.: Relationship of polyoma virus and tumor *in vivo*. Virology **12**, 463 (1960).

18. HABEL, K.: Resistance of polyoma virus immune animals to transplanted polyoma tumors. Proc. Soc. Exptl. Biol. Med. **106**, 722 (1961).

19. HABEL, K.: Immunological determinants of polyoma virus oncogenesis. J. Exptl. Med. **115**, 181 (1962).

20. HABEL, K.: Polyoma tumor antigen in cells transformed *in vitro* by polyoma virus. Virology **18**, 553 (1962).

21. HABEL, K.: Antigenic properties of cells transformed by polyoma virus. Cold Spring Harbor Symposia on Quant. Biol. **27**, 433 (1962).

22. HABEL, K.: The relationship between polyoma virus multiplication immunological competence, and resistance to tumor challenge in the mouse. Ann. N. Y. Acad. Sci. **101**, 173 (1962).

23. HABEL, K., and EDDY, B. E.: Specificity of resistance to tumor challenge of polyoma and SV40 virus-immune hamsters. Proc. Exptl. Biol. Med. **113**, 1 (1963).

24. HABEL, K.: Specific complement-fixing antigens in polyoma tumors and transformed cells. Virology **25**, 55 (1965).

25. HABEL, K., JENSEN, F., PAGANO, J. S., and KOPROWSKI, H.: Specific complement-fixing tumor antigen in SV40-transformed human cells. Proc. Soc. Exp. Biol. Med. **118**, 4 (1965).

26. HABEL, K.: Cancer Research, *to be published*.

27. HABEL, K.: Intern. Symposia Immunopathol., Monaco, in press (1965).

28. HAMBURG, V. P., and SVET-MOLDAVSKY, G. J.: Artificial heterogenization and tumours by means of Herpes Simplex and polyoma viruses. Nature (London) **203**, 772 (1964).

29. HANAFUSA, H., HANAFUSA, T., and RUBIN, H.: The defectiveness of Rous sarcoma virus. Proc. Natl. Acad. Sci. **49**, 572 (1963).

30. HARE, J. D.: Transplant immunity to polyoma virus induced tumors. I. Correlations with biological properties of virus strains. Proc. Soc. Exptl. Biol. Med. **115**, 805 (1964).

31. HARTLEY, J. W., ROWE, W. P., CAPPS, W. I., and HUEBNER, R. J.: Complement fixation and tissue culture assays for mouse leukemia viruses. Proc. Natl. Acad. Sci. **53**, 931 (1965).

32. HELLSTRÖM, I.: Distinction between the effects of antiviral and anticellular polyoma antibodies on polyoma tumor cells. Nature **208**, 652 (1965).
33. HUEBNER, R. J., ROWE, W. P., TURNER, H. C., and LANE, W. T.: Specific adenovirus complement-fixing antigens in virus-free hamster and rat tumors. Proc. Natl. Acad. Sci. **50**, 379 (1963).
34. HUEBNER, R. J., ARMSTRONG, D., OKUYAN, M., SARMA, P. S., and TURNER, H. C.: Specific complement-fixing viral antigens in hamster and guinea pig tumors induced by the Schmidt-Ruppin strain of avian sarcoma. Proc. Natl. Acad. Sci. **51**, 742 (1964).
35. HUEBNER, R. J., PEREIRA, H. G., ALLISON, A. C., HOLLINSHEAD, A. C., and TURNER, H. C.: Production of type-specific C antigen in virus-free hamster tumor cells induced by adenovirus type 12. Proc. Natl. Acad. Sci. **51**, 432 (1964).
36. HUEBNER, R. J., CASEY, M. J., CHANOCK, R. M., and SCHELL, K.: Tumors induced in hamsters by a strain of adenovirus type 3: sharing of tumor antigens and "neoantigens" with those produced by adenovirus type 7 tumors. Proc. Natl. Acad. Sci. **54**, 381 (1965).
37. JONSSON, N., and SJÖGREN, H. O.: Further studies on specific transplantation antigens in Rous sarcoma of mice. J. Exptl. Med. **122**, 403 (1965).
38. KHERA, K. S., ASHKENAZI, A., RAPP, F., and MELNICK, J. L.: Immunity in hamsters to cells transformed *in vitro* and *in vivo* by SV40. J. Immunol. **91**, 604 (1963).
39. KIT, S., FREARSON, P. M., and DUBBS, D. R.: Enzyme induction in polyoma-infected mouse embryo cells. Fed. Proc. **24**, 596 (1965).
40. KLEIN, G., SJÖGREN, H. O., KLEIN, E., and HELLSTRÖM, K. E.: Demonstration of resistance against methyl icholanthrene-induced sarcomas in the primary autochthonous host. Cancer Res. **20**, 1561 (1960).
41. KLEIN, E., and KLEIN, G. L.: Antigenic properties of lymphomas induced by the Moloney agent. J. Natl. Cancer Inst. **32**, 547 (1964).
42. KLEIN, G., SJÖGREN, H. O., and KLEIN, E.: Demonstration of host resistance against isotransplantation of lymphomas induced by the gross agent. Cancer Res. **22**, 955 (1962).
43. KOCH, M. A., and SABIN, A. B.: Specificity of virus-induced resistance to transplantation of polyoma and SV40 tumors in adult hamsters. Proc. Soc. Exptl. Biol. Med. **113**, 4 (1963).
44. LAW, L. W.: Neoplasms in thymectomized mice following room infection with polyoma virus. Nature **205**, 672 (1965).
45. LAW, L. W., and DAWE, C. J.: Influence of total body X-irradiation on tumor induction by parotid tumor agent in adult mice. Proc. Soc. Exptl. Biol. Med. **105**, 414 (1960).
46. LEVINTHAL, J. D., JAKOBOVITS, M., and EATON, M. D.: Polyoma disease and tumors in mice: the distribution of viral antigen detected by immunofluorescence. Virology **16**, 314 (1962).
47. NOYES, W. F., and MELLORS, R. C.: Fluorescent antibody detection of the antigens of the Shope papilloma virus in papillomas of the wild and domestic rabbit. J. Exp. Med. **106**, 555 (1957).
48. OLD, L. J., BOYSE, E. A., and STOCKERT, E.: Typing of mouse leukemias by serological methods. Nature **201**, 777 (1964).
49. OLD, L. J., and BOYSE, E. A.: Antigens of tumors and leukemias induced by viruses. Fed. Proc. **24**, 1009 (1965).
50. PASTERNAK, G., and GRAFFI, A.: Induction of resistance against isotransplantation of virus-induced myeloid leukemias. Brit. J. Cancer **17**, 532 (1963).
51. POPE, J. H., and ROWE, W. P.: Detection of specific antigen in SV40-transformed cells by immunofluorescence. J. Exptl. Med. **120**, 121 (1964).

52. RAPP, F., KITAHARA, T., BUTEL, J. S., and MELNICK, J. L.: Synthesis of SV40 tumor antigen during replication of simian papovavirus (SV40). Proc. Natl. Acad. Sci. **52,** 1138 (1964).

53. ROGERS, S.: Induction of arginase in rabbit epithelium by the Shope rabbit papilloma virus. Nature **183,** 1815 (1959).

54. ROWE, W. P., HARTLEY, J. W., LAW, L. W., and HUEBNER, R. J.: Studies of mouse polyoma virus infection. III. Distribution of antibodies in laboratory mouse colonies. J. Exptl. Med. **109,** 449 (1959).

55. ROWE, W. P.: The epidemiology of mouse polyoma virus infection. Bact. Rev. **25,** 18 (1961).

56. SABIN, A. B., SHEIN, H. M., KOCH, M. A., and ENDERS, J. F.: Specific complement-fixing tumor antigens in human cells morphologically transformed by SV40. Proc. Natl. Acad. Sci. **52,** 1316 (1964).

57. SABIN, A. B., and KOCH, M. A.: Source of genetic information for specific complement-fixing antigens in SV40 virus-induced tumors. Proc. Natl. Acad. Sci. **52,** 1131 (1964).

58. SACHS, L.: Transplantability of an x-ray-induced and a virus-induced leukemia in isologous mice inoculated with a leukemia virus. J. Natl. Cancer Inst. **29,** 759 (1962).

59. SJÖGREN, H. O., HELLSTRÖM, I., and KLEIN, G.: Transplantation of polyoma virus-induced tumor in mice. Cancer Res. **21,** 329 (1961).

60. SJÖGREN, H. O., HELLSTRÖM, I., and KLEIN, G.: Resistance of polyoma virus immunized mice to transplantation of established polyoma tumors. Exptl. Cell Res. **23,** 204 (1961).

61. SJÖGREN, H. O.: Studies on specific transplantation resistance to polyoma-virus-induced tumors. IV. Stability of the polyoma cell antigen. J. Natl. Cancer Inst. **32,** 661 (1964).

62. SJÖGREN, H. O.: Studies on specific transplantation resistance to polyoma-virus-induced tumors. I. Transplantation resistance induced by polyoma virus infection. J. Natl. Cancer Inst. **32,** 361 (1964).

63. SLETTENMARK-WAHREN, B., and KLEIN, E.: Cytotoxic and neutralization tests with serum and lymph node cells of isologous mice with induced resistance against gross lymphomas. Cancer Res. **22,** 947 (1962).

64. STÜCK, B., OLD, L. J., and BOYSE, E. A.: Occurrence of soluble antigen in the plasma of mice with virus-induced leukemia. Proc. Natl. Acad. Sci. **52,** 950 (1964).

65. STÜCK, B., OLD, L. J., and BOYSE, E. A.: Antigenic conversion of established leukaemias by an unrelated leukaemogenic virus. Nature **202,** 1016 (1964).

66. TAKEMOTO, K. K., MALMGREN, R. A., and HABEL, K.: Immunofluorescent demonstration of polyoma tumor antigen in lytic infection of mouse embryo cells. Virology, *in press* (1966).

67. Tevethia, S. S., KATZ, M., and RAPP, F.: New surface antigen in cells transformed by simian papovavirus SV40. Proc. Soc. Exptl. Biol. Med. **119,** 896 (1965).

68 TODARO, G. J., HABEL, K., and GREEN, H.: Antigenic and cultural properties of cells doubly transformed by polyoma virus and SV40. Virology **27,** 179 (1965).

69. TRENTIN, J. J., and BRYAN, E.: Immunization of hamsters and histoisogenic mice against transplantation of tumors induced by human adenovirus type 12. Proc. Am. Assoc. Cancer Res. **5,** 64 (1964).

70. VANDERPUTTE, M., DENYS, P., JR., LEYTEN, R., and DE SOMER, P.: The oncogenic activity of the polyoma virus in thymectomized rats. Life Sci. **7,** 475 (1963).

71. VOGT, M., and DULBECCO, R.: Studies on cells rendered neoplastic by polyoma virus the problem of the presence of virus-related materials. Virology **16,** 41 (1962).

72. VOGT, P. K., and LUYKX, N.: Observations on the surface of cells infected with Rous sarcoma virus. Virology **20,** 75 (1963).

73. VOGT, P. K., Sarma, P. S., and HUEBNER, R. J.: Presence of avian tumor virus group-specific antigen in nonproducing Rous sarcoma cells of the chicken. Virology 27, 233 (1965).
74. WAHREN, B.: Further studies of the immunological properties of Gross virus-induced lymphomas. Cancer Res. 24, 906 (1964).
75. WEISS, D. W., FAULKIN, L. J., and DEOME, K. B.: Acquisition of heightened resistance and susceptibility to spontaneous mouse mammary carcinomas in the original host. Cancer Res. 24, 732 (1964).
76. ZILBER, L. A.: Studies on tumor antigens. J. Natl. Cancer Inst. 18, 341 (1957).

Discussion

BY V. DEFENDI

Wistar Institute of Anatomy and Biology,
Philadelphia, Pennsylvania

The cellular antigens induced by tumor viruses that Dr. Habel has just illustrated, appear during and because of a specific interaction of the viral genome with the host cells. This we can state with unquestionable certainty. Whether their presence in virus-free cells indicates persistence of the viral genome or a fraction of it, we cannot state with equal certainty since attempts to obtain independent proofs of viral genome integration in the case of SV40 and polyoma virus have given, at best, equivocal results (2, 12, 13). The alternate hypothesis, that the antigens represent products of the cell genome whose synthesis is derepressed as a consequence of a specific interaction with the virus, appears improbable but has not yet been completely excluded. We have to remember that there are at least two well-documented cases in which antigens, present in some organs during embryonal life, disappear in the adult, to reappear again when a neoplastic process has occurred. One is the case of the antigen in rat liver hepatomas (1), the other is the antigen observed in colon carcinomas of man (7). These are therefore examples in which the expression of specific cellular antigens is modulated by differentiation and growth.

But if we accept that the transplantation and complement-fixing antigen are specified by the viral genome, these antigens become extremely important, not only as possible diagnostic tools but also because through them it is possible to analyze the various functions of the viral genome as related to its oncogenic activity.

TABLE I

PRODUCTS AND FUNCTIONS OF ONCOGENIC VIRUSES (DNA)

Products	Functions
Cf antigen	Activation of host DNA synthesis
Transp. antigen	Transformation *in vitro*
Enzymes(?)	Viral replication
Viral protein(s)	Oncogenesis *in vivo*

After infection of appropriate cultures *in vitro* or of animals with SV40 or polyoma virus, various changes occur in the cells, as illustrated in Table I, that may be subdivided into specific products and functions. Transformation *in vitro* should be qualified for the different parameters taken into consideration, thus one should properly talk of morphological or antigenic or cytological transformation. Neoplastic transformation, that is, the end results of all the various changes, may be considered one and the same process as the oncogenesis *in vivo* in a different environment (3). The products whose synthesis is induced by the virus may be identified as the two types of antigens, the proteins of the virion coat and possibly the enzymes related to the viral replication. It has not been yet clarified whether the enzymes that increase after infection are virus specific or modified cellular enzymes (5, 11).

Since products represent the expression of the viral genome or of part of it, it may be asked which products are related to which functions. A clear answer cannot be given, but suggestive relations can be established on the basis of the facts available.

The properties of the complement-fixing antigen (ICFA or T) are listed in Table II. Only some of the properties require further elucidation. The antigen is found in transformed

TABLE II

CF ANTIGEN EXPRESSION

It precedes viral synthesis.
It does not require cell DNA or viral DNA(?) synthesis.
It is present in all noninfectious transformed or tumor cells, but it may be present in *nonproducer*, nontransformed cells.
It is inactivated (UV or X-ray) at lower rate than infectivity.
It is induced by the SV40 genome fraction incorporated in the adenovirus capsid. (LLE-46)

---

cells long after they have ceased to produce infectious virus. Conversely, during primary infection of the culture, ICFA synthesis precedes that of the virus, and it can be observed in cells in which viral synthesis does not occur and that may or may not transform in following generations. What is more significant, ICFA is present in the same cells in which cellular DNA synthesis is activated, indicating that partial transcription of the viral genome is compatible with cell multiplication (10). The sensitivity to UV and X-ray inactivation of SV40 for the property to induce synthesis of ICFA, to transform cells *in vitro* and to produce tumors *in vivo*, is of the same order. All these properties are more resistant to inactivation by a factor of 2 to 3-fold than infectivity (4).

Finally, the property of inducing SV40 ICFA is endowed in the SV40 fraction that is incorporated in the adenovirus capsids of the LLE-46 virus (9). All these facts indicate that synthesis of the complement-fixing antigen is regulated only by a fraction of the viral genome and that associated with this fraction are the functions of activation of DNA synthesis, of transforming cells *in vitro* and of producing tumors *in vivo*. It has not been conclusively demonstrated that the induction of complement-fixing antigen is necessary for the expression of the virus oncogenic properties.

Much less is known about the transplantation antigen in terms of viral replication and of the fraction of the viral genome responsible for it (Table III). But in an indirect way, it can be demonstrated that the transcription of the whole viral genome is not necessary

TABLE III

TRANSPLANTATION ANTIGEN

Relationship to viral replication—unknown. Presumably not required.

Present in all noninfectious transformed and tumor cells (variable immunogenicity).

It is produced by the SV40 genome fraction incorporated in the adenovirus capsid. (LLE-46)

for induction of virus antigen. If SV40 or SV40 transformed cells that are virus-free are inoculated in hamsters 30–40 days after SV40 infection, the incidence of primary tumors is lower than in animals inoculated only with the virus at birth (6, 8). The basis of this protective effect is interpreted as being an active sensitization against the viral-induced transplantation antigen by the intercurrent injection of virus. In one experiment in which various litters of hamsters were inoculated at birth with SV40 and the animals were then challenged with SV40 or with SV40 transformed human cells or with the LLE-46 virus, it was found that LLE-46 was just as effective in inhibiting SV40 primary tumors (Fig. 1). It can then be concluded that the fraction of the SV40 genome incorporated into adeno-7 capsids is able to induce SV40 transplantation antigen in addition to the complement-fixing antigen. We then find that the functions pertinent to the oncogenic properties of SV40 and polyoma viruses are associated with only a few specific products. In other words, transcription of only part of the viral genome is sufficient for the induction of tumors. These conclu-

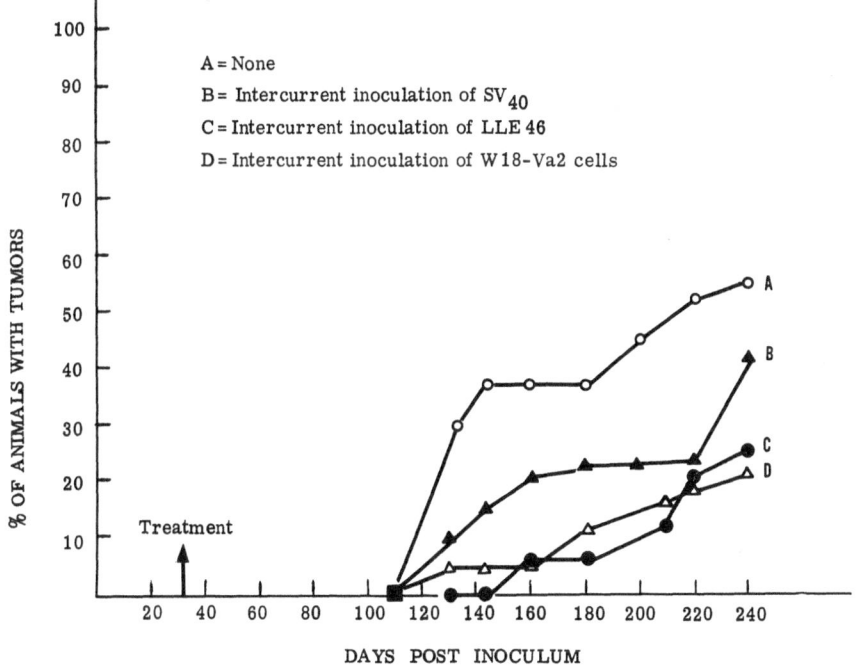

Fig. 1. Effect of various intercurring treatments on the induction of primary tumors by SV40.

sions may be carried even further, and it may be stated that the transcription must be incomplete for any of these viruses to be oncogenic, since if it were complete and the whole infectious virus were synthesized, the infected cells would die. An abortive infection therefore must be the basis of transformation *in vitro* and oncogenesis *in vivo* by SV40 and polyoma. The problem we have to resolve is whether the abortive infection is due to defective virus particles or to the ability of the infected cells to arrest the transcription of the viral genome at a certain specific point.

References

1. Abelev, G. I.: Study of the antigenic structure of tumors. Acta Unio contra Cancrum 19:80–82, 1963.
2. Axelrod, D., Habel, K., and Bolton, E. T.: Polyoma virus genetic material in a virus-free polyoma-induced tumor. Science 146:1466–1468, 1964.
3. Defendi, V.: Transformation *in vitro* of mammalian cells by polyoma and Simian 40 viruses. Progr. Exp. Tumor Res. 8: 1965 (*in press*), Karger, Basel.
4. Defendi, V., Jensen, F., and Sauer, G.: *Unpublished observations.*
5. Dulbecco, R., Hartwell, L. H., and Vogt, M.: Induction of cellular DNA synthesis by polyoma virus. Proc. Nat. Sci. (Washington) 53:403–410, 1965.
6. Eddy, B. E., Grubbs, G. E., and Young, R. D.: Tumor immunity in hamsters infected with adenovirus Type 11 or Simian virus 40. Proc. Soc. Exptl. Biol. Med. 117:575–579, 1964.
7. Gold, P., and Freedman, S. O.: Specific carcinoembryonic antigens of the human digestive system. J. Exp. Med. 122:467–481, 1965.
8. Goldner, H., Girardi, A. S., Larson, V. M., and Hilleman, M. R.: Interruption of SV40 virus tumorigenesis using irradiated homologous tumor antigen. Proc. Soc. Exptl. Biol. Med. 117:851–857, 1964.
9. Rowe, W. P., and Baum, S. G.: Evidence for a possible genetic hybrid between adenovirus Type 7 and SV40 viruses. Proc. Natl. Acad. Sci. (Washington) 52:1340–1347, 1964.
10. Sauer, G., and Defendi, V.: Stimulation of DNA synthesis and complement-fixing antigen production in human diploid cell cultures with SV40: evidence for abortive infection (*in preparation*).
11. Sheinin, L.: Studies on the thymidine kinase activity of mouse embryo cells infected with polyoma virus. Virology 28:47, 1966.
12. Winocour, E.: Attempts to detect an integrated polyoma genome by nucleic acid hybridization. I. "Reconstruction" experiments and complementarity tests between synthetic polyoma RNA and polyoma tumor DNA. Virology 25:276–288, 1965.
13. Winocour, E.: Attempts to detect an integrated polyoma genome by nucleic acid hybridization. II. Complementarity between polyoma virus DNA and normal mouse synthetic RNA. Virology 27:520–527, 1965.

Complementation Between Defective Oncogenic Viruses

By

Fred Rapp [1]

Department of Virology and Epidemiology,
Baylor University College of Medicine,
Houston, Texas

Complementation involves functional interaction between two viruses which results in replication under normally inhibitory conditions. A virus which requires complementation for replication is usually defective in some portion of its genome. However, some viruses competent to grow unilaterally in one cell may be incompetent in a second cell and thus require a second virus to furnish help for some step in the replicative cycle. Though complementation has been studied extensively with bacterial viruses (2, 20, 40, 43) and with one plant virus (15, 33), relatively few such studies have been carried out with animal viruses.

The best studied animal virus system involves the Bryan strain of Rous sarcoma virus which is now known to be defective and requires another member of the avian leucosis virus complex to supply a protein coat (13, 38). Apparently, synthesis of dermovaccinia virus in the mouse L cell also requires co-infection with neurovaccinia virus (23).

The demonstration that the oncogenic simian papovavirus SV40 could enhance the replication of adenovirus in green monkey kidney cells (21, 25) and that these two unrelated viruses had apparently formed a stable "hybrid" virus population unaffected by passage in the presence of antibody directed against SV40 (14, 31, 35) prompted the investigations described in this paper. The oncogenicity of the virus populations that have evolved during the course of this study and some of the significance and implications of the observations made will be discussed.

Potentiation of the Replication of Adenoviruses by SV40

A number of recent studies have revealed that many human adenoviruses replicate poorly, or not at all, in African green monkey kidney cells (1, 6, 8, 11, 12, 39). When the cells are co-infected with SV40, however, the adenoviruses replicate to

[1] The author wishes to thank his many collaborators for their contributions to the research described in this article. The support, advice, and encouragement of Dr. Joseph L. Melnick continues to catalyze the progress of these studies. This investigation was supported in part by Public Health Service Grants CA-04600 from the National Cancer Institute and AI 05382 from the National Institute of Allergy and Infectious Diseases, National Institutes of Health, United States Public Health Service.

TABLE I

EFFECT OF SV40 AND OTHER VIRUSES ON REPLICATION OF ADENOVIRUS
TYPES 2 AND 7 IN GREEN MONKEY KIDNEY CELLS

"Enhancing" virus	Adenovirus	Titer of adenovirus ($\log_{10}$ PFU/ml) hours post-inoculation			Fold enhancement at 72 hours
		1	24	72	
None	Type 7	4.7	3.9	3.8	—
SV40	Type 7	4.5	4.7	6.8	1000×
Herpes simplex virus	Type 7	—	—	4.1	2×
Measles virus	Type 7	—	—	3.7	None
Rabbit papilloma virus	Type 7	—	—	3.6	None
Human wart virus	Type 7	—	—	3.6	None
None	Type 2	4.9	4.5	4.0	—
SV40	Type 2	5.0	5.6	7.3	1800×

high titer (Table I). Similar results have been obtained with numerous types of human adenoviruses. Titers of SV40 are neither increased nor decreased in the presence of the adenoviruses. The requirement appears to be specific since other viruses cannot substitute for SV40 (Table I). The viruses tested include the papovaviruses, rabbit papilloma and human wart (which are not known to replicate in simian cells), herpes simplex virus (which replicates in the nucleus of these cells), and measles virus (which replicates in the cytoplasm of the cells).

Cells singly infected with adenoviruses, therefore, do not synthesize infectious virus or viral capsid antigens (11, 16). However, the cells do synthesize the adenovirus tumor (T) antigen following infection (Table II); deposits of this antigen

TABLE II

FORMATION OF ANTIGENS IN GREEN MONKEY KIDNEY CELLS INFECTED
WITH ADENOVIRUS IN THE PRESENCE AND ABSENCE OF SV40

Virus	Adenovirus-induced antigens		SV40-induced antigens	
	Tumor	Viral	Tumor	Viral
Adenovirus type 2	+	0	0	0
Adenovirus type 7	+	0	0	0
Adenovirus type 12	+	0	0	0
SV40	0	0	+	+
SV40 + Adenovirus types 2, 7, or 12	+	+	+	+

+ = Presence of antigen detectable by immunofluorescence.

0 = Absence of antigen detectable by immunofluorescence.

may correspond to the nuclear stippling previously observed in singly infected monkey cells (22).

It appears from these studies that the presence of SV40 is required in green monkey kidney cells for replication of human adenoviruses. Reasons why these adeno-viruses undergo an abortive cycle in the simian cells are unknown at the present time. The requirement for SV40 is specific, however, and the SV40 genome appears to contribute some information (perhaps the code for essential enzymes) that enables adenoviruses to replicate in cells derived from green monkey kidneys. The following section will explore in detail the interaction of a defective SV40 genome with adenoviruses, a phenomenon which enables both genomes to replicate and allows transfer of oncogenic potential from one adenovirus to a heterotypic adenovirus.

Adenovirus-Para (Defective SV40) Viruses

In the previous section, co-infection of GMK cells by adenoviruses and SV40 resulted in synthesis of both viruses. SV40, but not the adenovirus, was capable of replicating unilaterally. The discovery that a population of adenovirus type 7 was carrying SV40 determinants and that replication of the virus did not result in the synthesis of infectious SV40 suggested that these two unrelated viruses might have hybridized to form a stable, unique genome. The properties of this virus were, therefore, carefully scrutinized and the following sections summarize the information presently available.

General Properties

The original adenovirus involved was type 7 (strain L. L.) which had been isolated in 1955 at Fort Ord from military recruits. The virus had been carried through 22 serial passages in primary rhesus kidney cells and had been used as the type 7 prototype in commercially produced adenovirus vaccines. It appears that somewhere before the 9th serial passage in rhesus kidney cells, this strain was inadvertently contaminated with SV40; the 23rd and 24th passages, grown in green monkey kidney tissue cultures, were therefore treated with rabbit anti-serum against SV40. Subsequent passages in green monkey kidney cells were found to be free of infectious SV40.

The 28th passage of this virus (stock E46) and the 29th passage (stock SP2) induced tumors when inoculated into newborn hamsters (14) (Table III). Most of the hamsters carrying primary or transplanted tumors developed complement-fixing antibodies to SV40 T antigen and about 50% of the animals also developed antibodies to adenovirus T antigen (Table IV). Cells from the tumors were virus-free but synthesized SV40 T antigen (Table V). The same virus populations were also shown to induce the SV40 T antigen in tissue cultures (31, 34); both immuno-fluorescence and complement-fixation were used to detect the antigen. The SV40 T antigen could be detected following the inoculation of monkey, human, rabbit, and hamster cells; adenovirus penetration and replication seemed required for synthesis of the new antigen to take place since induction of the SV40 T antigen was directly related to the appearance of adenovirus cytopathic effects. At no time was SV40

Fred Rapp

TABLE III

Oncogenicity of SV40, Adeno 7-PARA and Adenovirus 7
in Newborn Hamsters

Virus	No. of animals inoculated	No. of animals with tumors *	Percent positive
SV40	89	54	61
Adeno 7-PARA	197	97	49
Adenovirus 7	158	0	0
SV40 + PARA-7	48	20	42

* Experiments terminated 9 months after inoculation.

TABLE IV

Analysis of Adeno 7-PARA Hamster Tumor Sera

Tests	Results	
	No. tested	No. positive
Reacted with homologous cells	19	16
Reacted with SV40 tumor antigen	12	11
Reacted with adenovirus 7 tumor antigen	45	22

TABLE V

Analysis of Adeno 7-PARA Tumors

Tests	Results	
	No. tested	No. positive
No. of tumors analyzed	22	
Cells grown in culture	22	19
Viruses isolated from cultured cells	19	0
Presence of virus particles	11	0
Presence of SV40 tumor antigen *	19	17
Presence of SV40 virus antigen *	19	0
Presence of adenovirus 7 viral antigen *	19	0

* By immunofluorescence.

capsid protein or infectious SV40 detected. Synthesis of the SV40 T antigen following inoculation of the "hybrid" into green monkey kidney cells (19) seemed to follow the same biochemical pathway as when SV40 itself was used to induce synthesis of the antigen (7, 28). In the electron microscope, adenovirus and adenosatellite particles were readily visualized (18) but no particles resembling SV40 could be detected. These findings were in direct contrast with the previous detection of SV40 particles and infectious virus in contaminated stocks of adenoviruses (44).

Neutralization of the ability to induce the SV40 T antigen was successful only when anti-adenovirus 7 serum was used. Sera against SV40 or against tumor antigen were without effect. These results strongly suggested that the SV40 determinants were encased in an adenovirus 7 capsid.

Recently, Black and Todaro (3) successfully used the virus to transform cells in cultures of primary hamster kidney and of diploid human skin. Serum against adenovirus type 7, but not against SV40, neutralized the transforming ability of the virus. The transformation observed resembled that induced by SV40, and almost all cells in the transformed cultures synthesized the SV40 T antigen. Attempts to detect infectious SV40, capsid antigens of SV40, or adenovirus type 7 tumor antigen were unsuccessful. These results further supported the concept that a portion of SV40 genome was encased in an adenocapsid, and that the SV40 determinants apparently include those required for induction of T antigen and for transformation.

Fate of SV40 Determinants in Simian and Human Cells

Rowe and Baum (35) noted that when terminal dilutions of the adenovirus 7-PARA (SV40) virus were carried out, the ability to induce the synthesis of SV40 T antigen was lost. Subsequently, techniques were developed for the plaque purification of the hybrid population. Plaque isolations were carried out in simian and human cells and the progeny were tested in both human embryonic and green monkey kidney cells for ability to induce SV40 T antigen. As can be seen in Table VI, isolates from green monkey kidney cells invariably retained the SV40 determinants but plaque isolates derived from human embryonic kidney cells had lost the ability

TABLE VI

INDUCTION OF SV40 TUMOR ANTIGEN BY PLAQUE ISOLATES
FROM AN ADENOVIRUS-PARA (SV40) VIRUS

		Number inducing tumor antigen	
Cells used for plating and passage	Number of isolates	Plaque isolates tested directly	Plaque isolates tested after 1 passage
Human kidney cells	60	0	0
Monkey kidney cells	17	13	17

to induce the synthesis of the T antigen (4, 35). These results could best be explained by the following hypotheses: (1) the particle carrying the SV40 determinant cannot replicate in human cells; (2) two particles are present with one replicating in human cells and with a second (the one carrying the SV40 determinants) replicating only in simian cells, or (3) a combination of hypotheses 1 and 2. These hypotheses were then tested experimentally.

Requirements for Plaque Formation in Simian and Human Cells

It had been noted that the titer of the adenovirus-PARA (SV40) virus was consistently higher when the virus was plaqued in human embryonic kidney cells than when it was plaqued in green monkey kidney cells. This finding, coupled with the observation that the determinants for SV40 are lost upon plaquing in human cells, suggested the possibility that two particles are required for plaque formation in the simian cells. This hypothesis turned out to be correct (5, 36).

It was first observed that the number of plaques formed following inoculation of simian cells by either the parent virus or plaque progeny derived from that virus did not directly follow the dilution factor. Numerous experiments revealed (as shown in the example charted in Fig. 1, left side) that the number of plaques formed approximated the square of the dilution, a finding consistent with the concept that two particles were required for plaque formation. These results were in contrast to

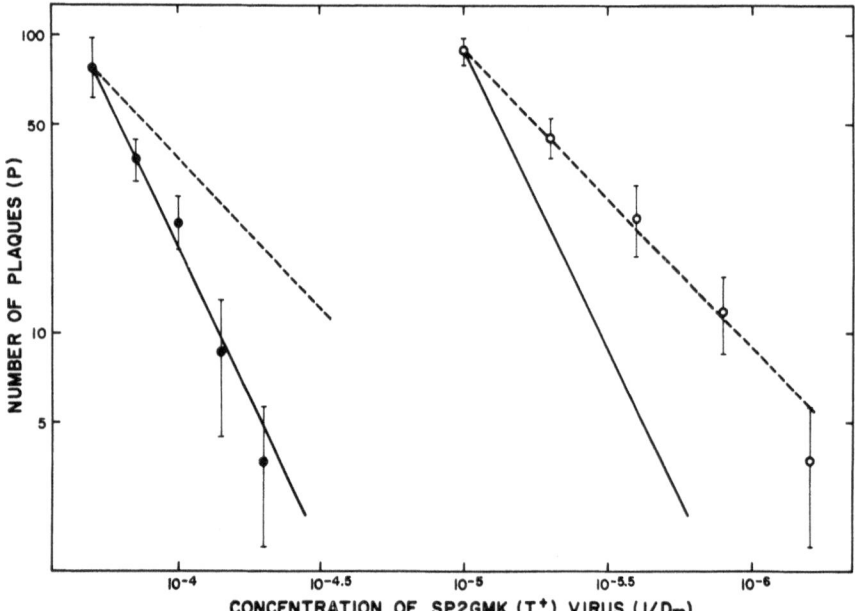

Fig. 1. Plaque counts obtained by plating PARA (SV40)-adenovirus, SP2 GMK (T+), on green monkey kidney cells in absence (filled circles) and presence (empty circles) of added helper adenovirus. Broken line is theoretical curve for one particle requirement. Solid line is theoretical curve for two particle requirement. The bars give the 95% confidence intervals.

those obtained when the virus was plaqued in human embryonic kidney cells, where there was a requirement for only one particle for plaque formation by the virus or progeny derived either from green monkey kidney or human embryonic kidney cells. Similar observations were made for the induction of cytopathic effects by the virus in tube cultures of simian cells. A detailed description of the data and their mathematical analysis can be found in the report by Boeyé, Melnick and Rapp (5).

Enhancement of Plaque Titers in Simian Cells

Since two particles appeared to be required for plaque formation in simian cells, it seemed reasonable to expect that the lower concentration of one of the particles might be limiting the number of plaques obtained. If true, saturation of the system would be achieved by the addition of this particle. This hypothesis was tested by adding the adenovirus that had been obtained by plaque-purification of the adenovirus-PARA (SV40) virus in human embryonic kidney cells. The virus had been shown to be free of SV40 determinants and represented one of the two particles present in the original population. The addition of the adenovirus at a multiplicity designed to cause infection of every cell in the monkey kidney culture was followed by varying dilutions of the PARA virus (as illustrated in Fig. 2); this procedure resulted in the enhancement of the plaque titer of the virus by a factor of more than 20-fold (Table VII). The helper adenovirus did not plaque in the absence of PARA (Fig. 2, far right plate). Formation of plaques by the virus on monkey

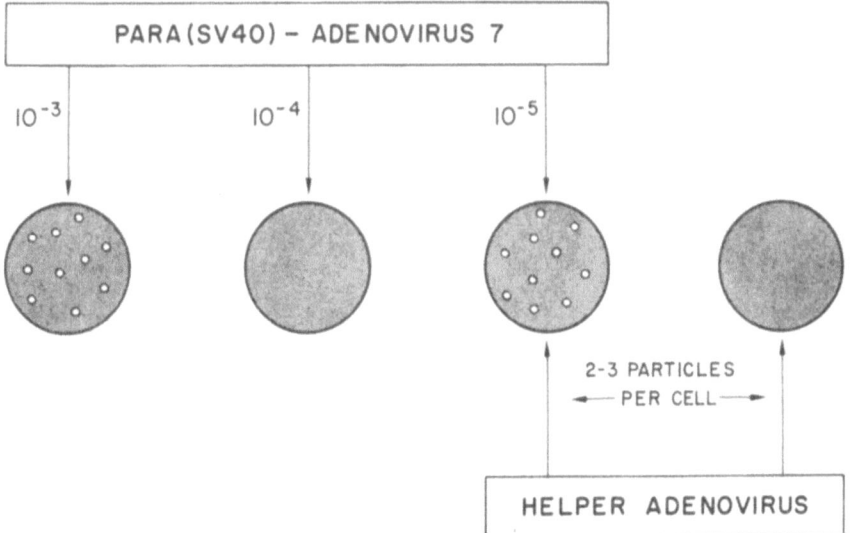

ENHANCEMENT OF PLAQUE TITER OF PARA POPULATIONS
BY HELPER ADENOVIRUSES

Fig. 2.

TABLE VII

ENHANCEMENT BY ADENOVIRUSES OF PLAQUE FORMATION BY PARA
(SV40)-ADENOVIRUS 7 IN GREEN MONKEY KIDNEY CELLS

Enhancing virus	PARA-adenovirus 7 titer ($\log_{10}$ PFU/ml)	Enhancement of PARA
None	4.9	—
Adenovirus type 7	6.4	30 fold
Adenovirus type 2	6.5	40 fold
Adenovirus type 12	6.7	60 fold

kidney cells all of which were infected with adenovirus, now followed a pattern consistent with the hypothesis that only one particle in the adenovirus-PARA (SV40) was required to cause the induction of the plaques observed (5); this is illustrated in Fig. 1 (right side). Rowe and Baum (36) have made similar observations.

As seen in Table VII, it was not necessary to use the homologous adenovirus type 7 as helper virus but adenovirus types 2 and 12 were equally effective (30, 34). An important corollary finding was that progeny from the terminal plaques formed under conditions of enhancement always carried the determinant of SV40 that allows induction of tumor antigen. This meant that the particle containing the SV40 determinant was required for plaque formation in green monkey cells and it, in turn, required the help of adenovirus to produce the cytopathic effects necessary for plaque formation. This was further confirmed by diluting the virus so that the multiplicity

TABLE VIII

PROPERTIES OF ADENOVIRUS, PARA,* AND SV40

Property	Designation of genetic markers	Adenovirus	PARA	PARA + adenovirus	SV40
Induction of SV40 T antigen	T	0	probably 0	+	+
Induction of SV40 V antigen	V	0	0	0	+
Induction of adeno 7 antigen	Ad$_7$	+	0	+	0
Replication in GMK cells	GMK	0	0	+	+
Replication in HEK cells	HEK	+	0	+	±
Presence of adenovirus capsid		+	+	+	0

T = tumor antigen; V = virus antigen; GMK = green monkey kidney cells; HEK = human embryonic kidney cells.

* PARA = beside, alongside of, amiss.

P = particle   A = aiding   R = replication of   A = adenovirus of infection involved only a single particle per cell. Under these conditions, no replication occurred (8). However, when adenoviruses were added as helpers to the dilute (non-infectious) PARA virus, newly synthesized adenovirus-PARA (SV40) virus was readily detected. This is the reason that the particle carrying the SV40 determinant was called PARA, which is an acronym for *p*article *a*iding and aided by *r*eplication of *a*denovirus (30). The properties of PARA, adenovirus and SV40 are summarized in Table VIII.

Boeyé, Melnick and Rapp (5) had calculated that the efficiency of the helper effect of the adenovirus in the parent adenovirus-PARA population could be accounted for only if the plating efficiency of the adenovirus was greater in the simian cells than in the human cells. Careful analysis of the helper phenomenon by Butel, Melnick and Rapp (6) revealed that the amount of helper adenovirus required per cell was less than one plaque-forming unit (as measured on human embryonic kidney cells, the most sensitive system available). Only 3 to 9 physical particles were required (Fig. 3); this corresponded to 0.02 to 0.14 PFU per cell (Fig. 4). Adenovirus inactivated by heat or ultraviolet irradiation was ineffective. When the helper virus was purified by equilibrium sedimentation in cesium chloride, the helper activity remained with the fraction (density of 1.34) containing the majority of infectious particles (Fig. 5). This study revealed, therefore, that almost all the

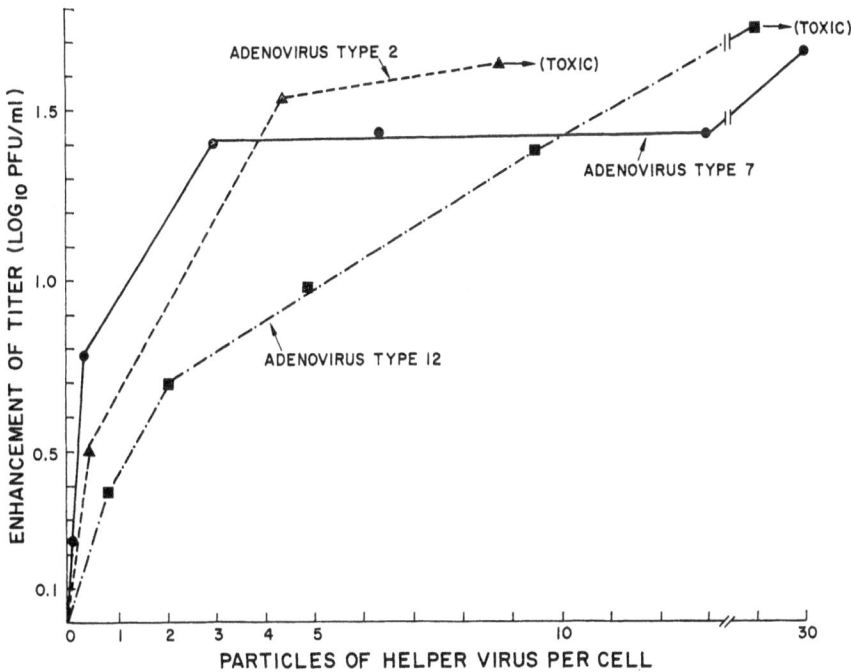

Fig. 3. Number of particles of helper adenovirus required per cell to enhance the titer of PARA (SV40)-adenovirus 7 in green monkey cells.

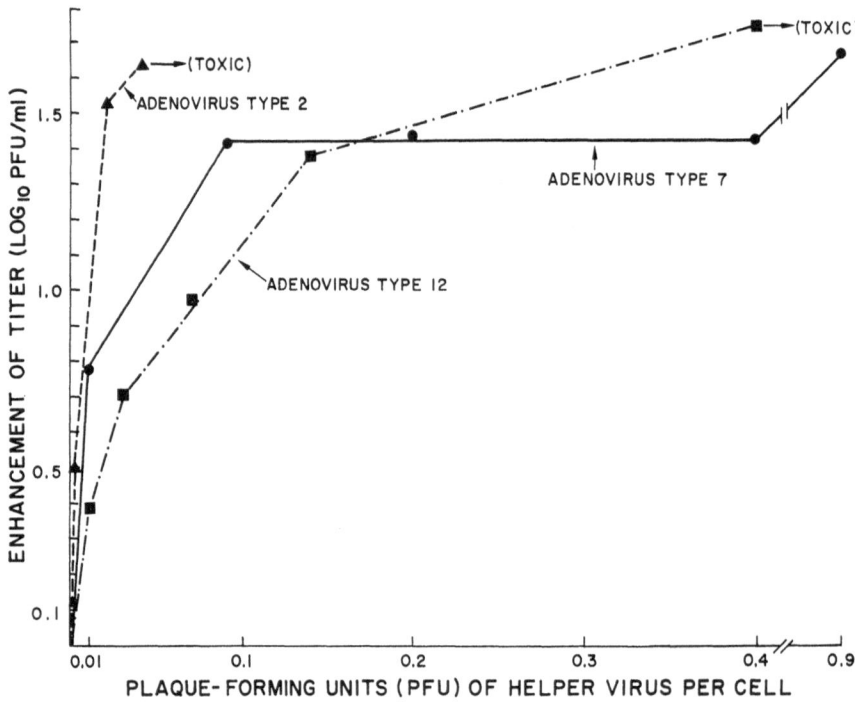

Fig. 4. Number of plaque-forming units (PFU) of helper adenovirus required per cell to enhance the titer of PARA (SV40)-adenovirus 7 in green monkey cells.

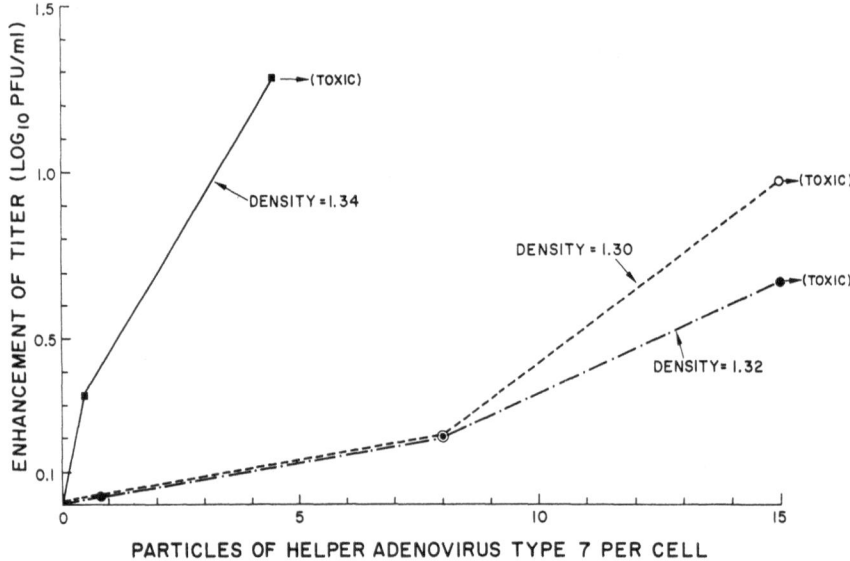

Fig. 5. The ability of adenovirus particles with various densities to enhance plaque formation in green money kidney cells by a PARA (SV40)-adenovirus.

adenovirions in the stocks tested were capable of enhancing the replication of PARA and could concurrently replicate their own progeny in the simian cells.

Transcapsidation

The mutual requirement of PARA and adenovirus for replication in green monkey cells and the finding that a number of adenoviruses would substitute as helpers reopened the question of the nature of the capsid of the PARA virus. Consequently, plaque isolates were obtained from plaques enhanced by various helper adenoviruses and the progeny characterized for ability to induce the SV40 T antigen as well as the ability to aid helper adenoviruses to plaque in green monkey kidney cells (29, 34). Table IX gives some representative results. It can be seen that all progeny treated with the tris saline (control) contained the ability to induce SV40 T antigen. However, the ability of the progeny to induce this antigen was generally neutralized with serum that had been prepared against the helper adenovirus. These results indicated that PARA was now encased in a capsid similar to that of the helper adenovirus and not in the original type 7 capsid. Similar results were obtained when the progeny were tested for ability to plaque in the presence of helper adenoviruses. If the original type 7 adenovirus PARA was still encased in an adenovirus 7 protein coat, only adenovirus 7 immune serum should neutralize it. Failure of the serum to do so suggested an antigenic shift in the capsid enclosing the SV40 determinants. These results revealed that the genetic determinants of one virus (SV40) can be carried in the capsid of an unrelated virus (adenovirus); as described above, such a population breeds true. The capsid can be changed, however, by the addition of a heterotypic adenovirus during the replication cycle. This phenomenon was called transcapsidation (30).

It now appears obvious that the adenovirus in the population replicates in human embryonic kidney cells by itself but replicates in green monkey kidney cells only in the presence of PARA. Whether PARA can replicate in the human cells remains unknown although results so far available suggest that it cannot do so as efficiently as it does in the green monkey cells. The results described have been

TABLE IX

NEUTRALIZATION * OF PROGENY FROM ADENOVIRUS-ENHANCED SP2 PLAQUES

Enhancing virus	SP2 Plaque Progeny Neutralized with:			
	Tris	Anti-adeno 7	Anti-adeno 2	Anti-adeno 12
Adenovirus type 7	0/9	9/9	Not done	Not done
Adenovirus type 2	0/12	0/12	11/12	Not done
Adenovirus type 12	0/14	3/14	Not done	11/14

* Of ability to induce synthesis of SV40 T antigen in GMK cells.

concerned primarily with an adenovirus type 7-PARA (SV40). However, the transcapsidation procedures suggest that any adenovirus can carry the SV40 determinants for the induction of tumor antigen; some or all of such virus populations may then be oncogenic (17).

Transplantation Antigens

Cells transformed by SV40 synthesize the SV40 T antigen (24, 29). Such cells also possess new surface antigens possibly associated with transformation (26, 41, 42). These findings raised the possibility that the SV40 determinants carried by PARA might include transplantation rejection antigens. One way to test this hypothesis was to inoculate weanling hamsters with the hybrid and then to challenge the animals with cells transformed by SV40. Such experiments revealed that this procedure confers resistance and that this resistance is comparable to that afforded by SV40 itself (32). Representative experiments are tabulated in Table X. The number of cells required to induce tumors following inoculation of the hybrid was considerably higher than the number required for animals inoculated with adenovirus 7 or with other control materials. The results indicate that incorporated in the PARA (SV40)-adenovirus is the SV40 determinant responsible not only for inducing the intranuclear T antigen but also the determinant responsible for conferring transplantation resistance.

TABLE X

SUMMARY OF TPD_{50} FOR SV40-TRANSFORMED CELLS (H-50)
FOLLOWING VACCINATION OF HAMSTERS WITH DIFFERENT
VIRUS POPULATIONS

Treatment	TPD_{50} in weeks			
	6	8	10	12
SV40	$>10^{5.0}$	$>10^{5.0}$	$>10^{5.0}$	$>10^{5.0}$
PARA (SV40)-				
Adenovirus 7	$>10^{5.0}$	$>10^{5.0}$	$10^{4.6}$	$10^{4.6}$
Adenovirus 7	$10^{4.5}$	$10^{3.5}$	$10^{2.7}$	$10^{2.4}$
Non-viral controls	$10^{4.5}$	$10^{3.5}$	$10^{3.2}$	$10^{3.0}$

Replication of PARA

Using the procedures described in the previous sections, it was now possible to measure separately the replication of the adenovirion and of the PARA virus in simian cells (8). PARA replication could be studied by plating harvests from one-step growth curves onto monkey kidney cells under conditions in which each cell was co-infected by a helper virus. The replication of adenovirus in the hybrid population resembled that for adenovirus grown in the presence of SV40 (Fig. 6).

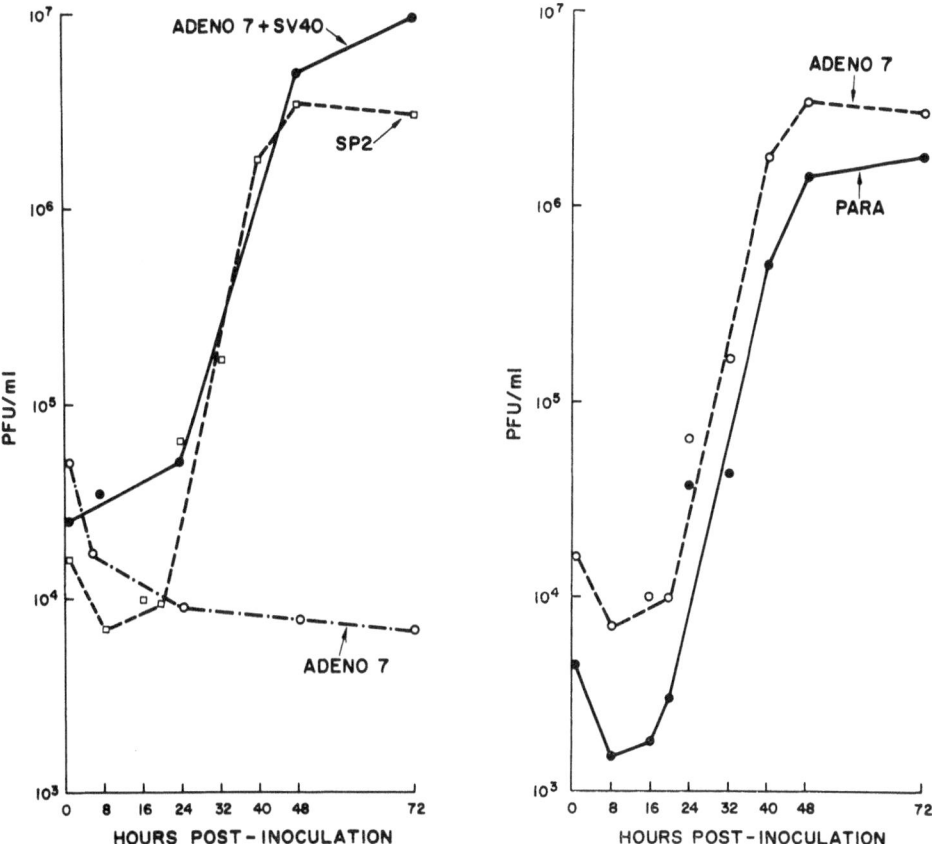

Fig. 6. Replication of PARA (SV40)-adenovirus 7 (SP2 stock) and of adenovirus type 7 (in the presence and absence of SV40) in green monkey kidney cells.

Fig. 7. Replication of adenovirus type 7 and PARA in green monkey kidney cells.

As previously noted, the adenovirus did not replicate at all in the absence of SV40. A latent period of 16 to 20 hours was followed by rapid increase in infectious virus; maximum titers were obtained 48 hours following inoculation of the monkey cells. The one-step growth curve for PARA was almost identical with that of the adenovirus (Fig. 7). Latent periods were similar, rate of increase in infectious particles coincided closely, and maximum titers were obtained at 48 hours (8). Most of the virus (both PARA and the adenovirus) remained cell-associated (27). The titration of the PARA-adenovirus 7 revealed that the ratio of infectious units of both components in the population was fairly constant and that the adenovirus outnumbered PARA by 2–4 to 1.

When adenovirus type 2 or type 12-PARA populations were studied, it was observed that the number of infectious adenovirions to infectious PARA particles was very high (27). However, most of the virus remained cell-associated and the growth curves (Figs. 8, 9) followed the pattern exhibited by the adenovirus 7-PARA population.

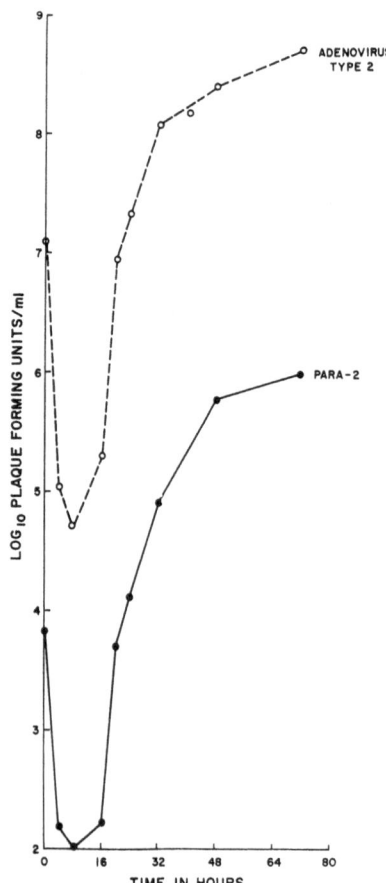

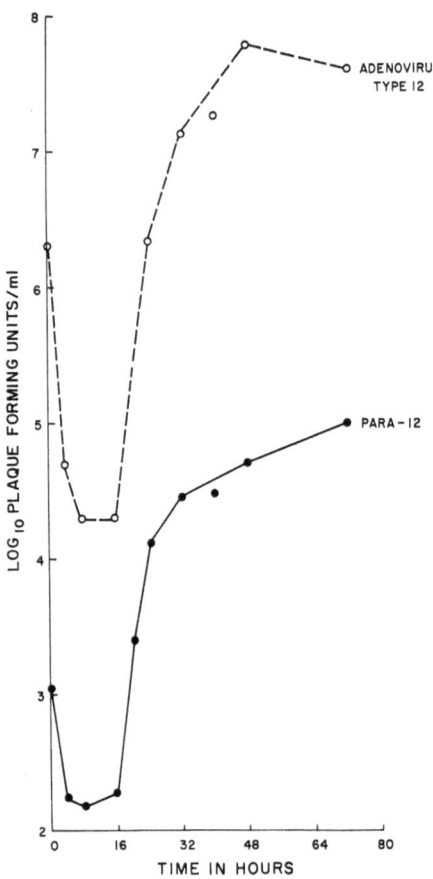

Fig. 8. Replication of adenovirus type 2 and PARA in green monkey kidney cells.

Fig. 9. Replication of adenovirus type 12 and PARA in green monkey kidney cells.

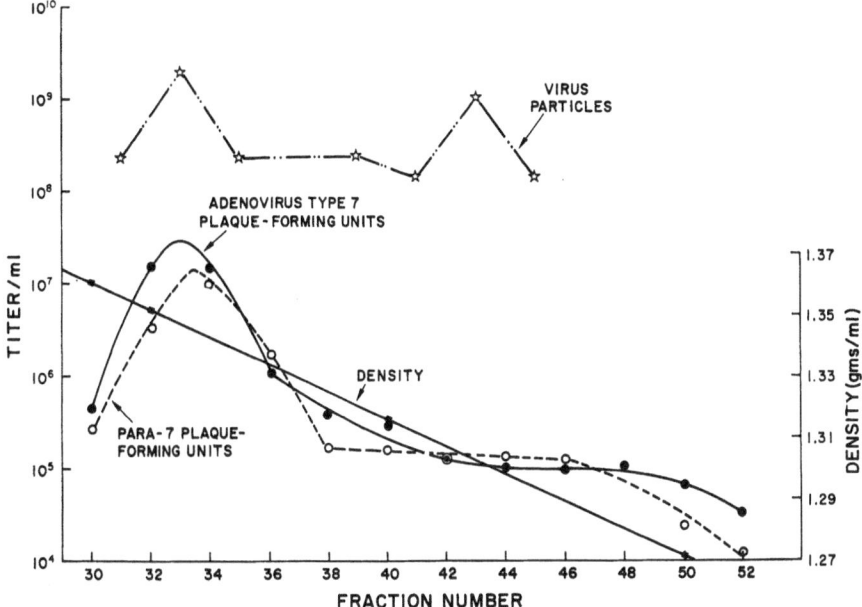

Fig. 10. Characterization of PARA and adenovirus by equilibrium centrifugation in cesium chloride.

Characterization of the PARA Particle

Attempts to separate the adenovirus and PARA by physical means have been unsuccessful. Physical procedures such as banding in cesium chloride and separation by filtration procedures have failed (27, 37). Figure 10 depicts such an attempt. The infectivity of PARA and of the adenovirus band together in cesium chloride at a buoyant density of about 1.34; a second peak of particles consists primarily of empty, and therefore non-infectious, units. Attempts to thermally inactivate one of the two components have also failed (5, 27).

Discussion

The interaction of two unrelated viruses for replication and transformation is a phenomenon unique in animal virology. Such interactions can apparently serve to alter the spectrum for virus replication, enhance oncogenic potential, and provide additional chances for genetic interaction, thereby increasing the survival potential of the virus determinants involved. These facts are manifest in the ability of the adenovirus-PARA population to replicate in simian cells, a property neither of the two particles in the population has in the absence of its partner.

Transcapsidation of the protein coat of PARA can apparently be accomplished with many adenoviruses. Preliminary information suggests that the resulting virus population is then oncogenic. It is possible that the oncogenicity is due to the introduction of the SV40 determinants into the host cells. The recent finding (39) that populations of adenovirus type 12 carrying SV40 determinants are more oncogenic than either parent virus suggests either a synergistic action of the two genomes leading to transformation or the more efficient introduction of the SV40 determinants into hamster cells when the genes are in an adenocapsid. The latter theory may be the more plausible since evidence is accumulating that adenoviruses can induce tumor antigens (thereby indicating that the virus adsorbs, penetrates and uncoats) in a high percent of hamster cells exposed to the virus while SV40 (in an SV40 capsid) can induce its tumor antigen in only a very small percent of cells derived from hamster tissues. The systems described in this paper probably represent only the forerunner of other examples of cooperation between DNA viruses. Easton and Hiatt (9, 10) and Beardmore and his colleagues (1) have already described results which suggest that the full SV40 genome can be carried in an adenovirus type 4 capsid.

The above studies were made possible by the development of methods and reagents which enabled the detection of the SV40 tumor antigen. Undoubtedly, situations exist for which markers are not available and it is possible that oncogenic potential of many tumor viruses rests in the defective genomes they may carry. For this reason, the search for such determinants must continue and it appears certain that the various paths taken by scientists entering this new area of research will lead to many exciting developments in viral oncology and molecular biology.

References

1. BEARDMORE, W. B., HAVLICK, M. J., SERAFINI, A., and McLEAN, I. W., JR.: Interrelationship of Adenovirus (Type 4) and Papovavirus (SV40) in Monkey Kidney Cell Cultures. *J. Immunol.* 95, 422–435 (1965).
2. BENZER, S.: Fine Structure of a Genetic Region in Bacteriophage. *Proc. Nat. Acad. Sci. U.S.* 41, 344–354 (1955).
3. BLACK, P. H., and TODARO, G. J.: In vitro Transformation of Hamster and Human Cells with the Adeno 7-SV40 Hybrid Virus. *Proc. Nat. Acad. Sci. U.S.* 54, 374–381 (1965).
4. BOEYÉ, A., MELNICK, J. L., and RAPP, F.: Adenovirus-SV40 "Hybrids": Plaque Purification into Lines in which the Determinant of the SV40 Tumor Antigen is Lost or Retained. *Virology,* 26, 511–512 (1965).
5. BOEYÉ, A., MELNICK, J. L., and RAPP, F.: SV40-Adenovirus "Hybrids"; Presence of Two Genotypes and the Requirement of Their Complementation for Viral Replication. *Virology,* 28, 56–70 (1966).
6. BUTEL, J. S., MELNICK, J. L., and RAPP, F.: Quantitative Aspects of Complementation Between Adenoviruses and a PARA (Defective SV40)-Adenovirus "Hybrid" Population. Submitted for publication, 1966.
7. BUTEL, J. S., and RAPP, F.: The Effect of Arabinofuranosylcytosine on the Growth Cycle of Simian Virus 40. *Virology,* 27, 490–495 (1965).
8. BUTEL, J. S., and RAPP, F.: Replication in Simian Cells of Defective Viruses in an SV40-Adenovirus "Hybrid" Population. *J. Bact.* 91, 278–284 (1966).
9. EASTON, J. M., and HIATT, C. W.: Possible Incorporation of SV40 Genome Within Capsid Proteins of Adenovirus 4. *Proc. Nat. Acad. Sci. U.S.* 54, 1100–1104 (1965).

10. EASTON, J. M., and HIATT, C. W.: Simian Virus 40: Replication in the Presence of Specific Antiserum and Adenovirus 4. *Science,* **151,** 582–583 (1966).

11. FELDMAN, L. A., BUTEL, J. S., and RAPP, F.: Interaction of Papovavirus SV40 and Adenoviruses. I. Induction of Adenovirus Tumor Antigen During Abortive Infection of Simian Cells. *J. Bact.,* in press, 1966.

12. FELDMAN, L. A., MELNICK, J. L., and RAPP, F.: Influence of SV40 Genome on the Replication of an Adenovirus-SV40 "Hybrid" Population. *J. Bact.* **90,** 778–782 (1965).

13. HANAFUSA, H., HANAFUSA, T., and RUBIN, H.: Analysis of the Defectiveness of Rous Sarcoma Virus. II. Specification of RSV Antigenicity by Helper Virus. *Proc. Nat. Acad. Sci. U.S.* **51,** 41–48 (1964).

14. HUEBNER, R. J., CHANOCK, R. M., RUBIN, B. A., and CASEY, M. J.: Induction by Adenovirus Type 7 of Tumors in Hamsters Having the Antigenic Characteristics of SV40 Virus. *Proc. Nat. Acad. Sci. U.S.* **52,** 1333–1340 (1964).

15. KASSANIS, B., and NIXON, H. L.: Activation of One Tobacco Necrosis Virus by Another. *J. Gen. Microbiol.* **25,** 459–471 (1961).

16. MALMGREN, R. A., RABSON, A. S., CARNEY, P. G., and PAUL, F. J. Immunofluorescence of Green Monkey Kidney Cells Infected with Adenovirus 12 and Adenovirus 12 Plus Simian Virus 40. *J. Bact.* **91,** 262–265 (1966).

17. MELNICK, J. L., RAPP, F., and BUTEL, J. S.: Oncogenicity of PARA (Defective SV40)-Adenovirus. *Proc. Am. Assoc. Cancer Res.,* in press (1966).

18. MELNICK, J. L., MAYOR, H. D., SMITH, K. O., and RAPP, F.: Association of 20-Millimicron Particles with Adenoviruses. *J. Bact.* **90,** 271–274 (1965).

19. MELNICK, J. L., and RAPP, F.: Effect of Inhibitors on Induction of SV40 Tumor Antigen by an Adenovirus-SV40 Hybrid. *Arch. f. d. ges. Virusforsch.* **17,** 424–435 (1965).

20. NISHIHARA, M., and ROMIG, W. R.: Temperature-Sensitive Mutants of *Bacillus subtilis* Bacteriophage SP3. II. *In vivo* Complementation Studies. *J. Bact.* **88,** 1230–1239 (1964).

21. O'CONOR, G. T., RABSON, A. S., BEREZESKY, I. K., and PAUL, F. J.: Mixed Infection with Simian Virus 40 and Adenovirus 12. *J. Nat. Cancer Inst.* **31,** 903–917 (1963).

22. O'CONOR, G. T., RABSON, A. S., MALMGREN, R. A., BEREZESKY, I. K., and PAUL, F. J.: Morphologic Observations of Green Monkey Kidney Cells after Single and Double Infection with Adenovirus 12 and Simian Virus 40. *J. Nat. Cancer Inst.* **34,** 679–693 (1965).

23. ODA, M.: Rescue of Dermovaccinia Abortive Infection by Neurovaccinia in L Cells. *Virology,* **25,** 665–666 (1965).

24. POPE, J. H., and ROWE, W. P.: Detection of Specific Antigen in SV40–Transformed Cells by Immunofluorescence. *J. Exp. Med.* **120,** 121–128 (1964).

25. RABSON, A. S., O'CONOR, G. T., BEREZESKY, I. K., and PAUL, F. J.: Enhancement of Adenovirus Growth in African Green Monkey Kidney Cell Cultures by SV40. *Proc. Soc. Exp. Biol. Med.* **116,** 187–190 (1964).

26. RAPP, F.: New Surface Antigens in Cells Transformed by Papovavirus SV40 and Preparation of Specific Cytotoxic Antibody. *Nat. Cancer Inst. Monograph,* in press (1965).

27. RAPP, F. (1966). Unpublished experiments.

28. RAPP, F., BUTEL, J. S., FELDMAN, L. A., KITAHARA, T., and MELNICK, J. L.: Differential Effects of Inhibitors on the Steps Leading to the Formation of SV40 Tumor and Virus Antigens. *J. Exp. Med.* **121,** 935–944 (1965).

29. RAPP, F., BUTEL, J. S., and MELNICK, J. L.: Virus-Induced Intranuclear Antigens in Cells Transformed by Papovavirus SV40. *Proc. Soc. Exp. Biol. Med.* **116,** 1131–1135 (1964).

30. RAPP, F., BUTEL, J. S., and MELNICK, J. L.: SV40-Adenovirus "Hybrid" Populations: Transfer of SV40 Determinants from One Type of Adenovirus to Another. *Proc. Nat. Acad. Sci. U.S.* **54**, 717–724 (1965).

31. RAPP, F., MELNICK, J. L., BUTEL, J. S., and KITAHARA, T.: The Incorporation of SV40 Genetic Material into Adenovirus 7 as Measured by Intranuclear Synthesis of SV40 Tumor Antigen. *Proc. Nat. Acad. Sci. U.S.* **52**, 1348–1352 (1964).

32. RAPP, F., Tevethia, S. S., and MELNICK, J. L.: Papovavirus SV40 Transplantation Immunity Conferred by an Adenovirus-SV40 Hybrid. *J. Nat. Cancer Inst.*, in press (1966).

33. REICHMANN, M. E.: The Satellite Tobacco Necrosis Virus: A Single Protein and Its Genetic Code. *Proc. Nat. Acad. Sci. U.S.* **52**, 1009–1017 (1964).

34. ROWE, W. P.: Studies of Adenovirus-SV40 Hybrid Viruses. III. Transfer of SV40 Gene Between Adenovirus Types. *Proc. Nat. Acad. Sci. U.S.* **54**, 711–717 (1965).

35. ROWE, W. P., and BAUM, S. G.: Evidence for a Possible Genetic Hybrid Between Adenovirus Type 7 and SV40 Viruses. *Proc. Nat. Acad. Sci. U.S.* **52**, 1340–1347 (1964).

36. ROWE, W. P., and BAUM, S. G.: Studies of Adenovirus-SV40 Hybrid Viruses. II. Defectiveness of the Hybrid Particles. *J. Exp. Med.* **122**, 955–966 (1965).

37. ROWE, W. P., BAUM, S. G., PUGH, W. E., and HOGGAN, M. D.: Studies of Adenovirus-SV40 Hybrid Viruses. I. Assay System and Further Evidence for Hybridization. *J. Exp. Med.* **122**, 943–954 (1965).

38. RUBIN, H.: Virus Defectiveness and Cell Transformation in the Rous Sarcoma. *J. Cell. Comp. Physiol.* **64**, 173–180 (1964).

39. SCHELL, K., LANE, W. T., CASEY, M. J., and HUEBNER, R. J.: Potentiation of Oncogenecity of Adenovirus Type 12 Grown in African Green Monkey Kidney Cell Cultures Preinfected with SV40 Virus: Persistence of Both T Antigens in the Tumors and Evidence for Possible Hybridization. *Proc. Nat. Acad. Sci. U.S.* **55**, 81–88 (1966).

40. TESSMAN, E. S.: Complementation Groups in Phage S13. *Virology*, **25**, 303–321 (1965).

41. TEVETHIA, S. S., KATZ, M., and RAPP, F.: New Surface Antigens in Cells Transformed by Simian Papovavirus SV40. *Proc. Soc. Exp. Biol. Med.* **119**, 896–901 (1965).

42. TEVETHIA, S. S., and RAPP, F.: Demonstration of New Surface Antigens in Cells Transformed by Papovavirus SV40 by Cytotoxic Tests. *Proc. Soc. Exp. Biol. Med.* **120**, 455–458 (1965).

43. VALENTINE, R. C., ENGELHARDT, D. L., and ZINDER, N. D.: Host-Dependent Mutants of the Bacteriophage f2. II. Rescue and Complementation of Mutants. *Virology*, **23**, 159–163 (1964).

44. YANG, C.-s., and MELNICK, J. L.: Contamination of Adenovirus Stocks with SV40 (Papovavirus Group). *Proc. Soc. Exp. Biol. Med.* **113**, 339–343. (1963).

Studies on Transformation by the Adenovirus-SV40 Hybrid Viruses

By

PAUL H. BLACK [1] AND HOWARD IGEL [2]

The adenovirus-SV40 hybrid viruses have been utilized for *in vitro* transformation experiments in order to answer the following questions:

1) Do all hybrids have the ability to transform primary weanling hamster kidney tissue culture cells?

2) What contribution to the transformation does the SV40 genome make, and would the effect on differentiation of hamster kidney cells observed with SV40 virus be seen by the hybrid viruses which contain only a portion of the SV40 genome?

3) What other markers of the SV40 genome are present in transformed cells other than the SV40 tumor (T) antigen?

4) Can a non-oncogenic adenovirus become oncogenic by the presence of a portion of the SV40 genome integrated with the adenovirus DNA?

5) Can enhancement of the oncogenic potential of an adenovirus be effected by the integration of a part of the SV40 genome?

6) If non-oncogenic adenoviruses, by hybridization, become oncogenic, would the tumors have a T antigen charactetristic of the non-oncogenic adenovirus?

7) What contributions to the transformation do the adeno and SV40 genomes make as regards morphology of the transformed cells?

Experimental Design

Primary weanling hamster kidney cells in roller tube cultures were infected with various hybrid viruses and were observed twice weekly for the appearance of transformation (1). When transformation was well advanced the cultures were dispersed with trypsin and passed. The cell lines were examined for morphology, and for antigen content by complement fixation (CF) and fluorescent antibody (FA) procedures, and were transplanted subcutaneously to hamsters. The resulting tumors were examined for antigen content by CF tests and were studied histologically; the

---

[1] Department of Health, Education and Welfare, Public Health Service, National Institutes of Health, National Institute of Allergy and Infectious Diseases, Laboratory of Infectious Diseases, Bethesda, Maryland.

[2] Department of Health, Education and Welfare, Public Health Service, National Institutes of Health, National Cancer Institute, Carcinogenesis Studies Branch, Bethesda, Maryland.

serum from tumor bearing hamsters was examined for CF antibodies. Much of this work is still in progress; thus only very preliminary data, frequently incomplete, will be presented.

Transformation in Weanling Hamster (WHK) Cultures

In Table I the results of experiments using the E46 pool of adenovirus 7 strain L.L. are given (1). The origin and history of this strain of virus have been described in detail (2). Briefly, the hybrid virus was produced by passage of adenovirus 7 in rhesus monkey kidney cultures which, unknowingly, contained SV40. When this latter fact was discovered, SV40 antiserum was then added and thereafter infectious SV40 disappeared from the passage line of virus. The E46 preparation, used in these experiments, contains 2 particles: the hybrid particle composed of portions of the adenovirus and SV40 genomes, and adenovirus 7 virions. The former particle is defective and needs the complete adeno 7 virion to replicate in African green monkey kidney (AGMK) cultures (3, 4, 5). The E46 pool used in these experiments will be referred to as Ad. 7^+. The data presented show that the Ad. 7^+ pool transforms WHK cultures at 8–26 days whereas no transformation occurred with the non-hybrid adeno 7 pool derived from it (6). Addition of SV40 antiserum had no effect on the transformation, whereas adenovirus 7 antiserum prevented the transformation. These findings are consistent with the data previously presented and indicate that the portion of the SV40 genome responsible for T antigen formation and neoplastic activity is enclosed in an adeno 7 protein coat (3, 6, 7).

In Table II, the results of experiments using other hybrid viruses are presented. The adenovirus 3 hybrid preparation originated in the same way as the E46 preparation; it contains 2 similar particles, one containing portions of both the adeno 3 and SV40 genomes while the second particle is a complete adeno 3 virion. This pool will be referred to as Ad. 3^+. The Ad. 2^{++} R 2TILB pool originated similarly. However, in addition to the hybrid and complete adeno 2 particles, this preparation contains free SV40 and a fourth particle which is composed of the SV40 genome in an adenovirus 2 protein coat (transcapsid particle) (Lewis, Prigge and Rowe, unpublished); it will be referred to as Ad. 2^{++}. The adeno 2^+ and 12^+ hybrid viruses were produced by growing the Ad. 7^+ (E46) preparations with adenoviruses 2 and 12, respectively, and then adding adeno 7 antiserum. In this way, the portion of the SV40 genome contained within the adeno 7 hybrid particle was transferred to the adenoviruses 2 and 12 (8, 9).

From the data in Table II, we note that the adeno 3^+ pool transformed WHK cultures while adeno 3 did not. The adeno 2^+ and 2^{++} preparations both transformed while adeno 2 did not. The role of the hybrid in the transformation by Ad. 2^{++} is unknown since free SV40 and the SV40 genome contained in an adeno 2 protein coat were present. The SV40 antiserum presumably was effective in neutralizing the infectious SV40 present and any SV40 virus produced after one infectious cycle due to infection with the transcapsid particle. Thus some cells probably received the complete SV40 genome brought in by the transcapsid particle

TABLE I

TRANSFORMATION EXPERIMENTS WITH THE ADENO 7-SV40 HYBRID VIRUS (AD. 7$^+$)

Tissue	Expt. no.	Inoculum	TCID$_{50}$/ 0.1 ml	First day of appearance of giant cells	Frequency of transformation, no. tubes pos./ no. tubes inoc.	Day transformation first observed in positive tubes	Length of observation of neg. tubes
WHK	11–25	LLE46 AG8	10$^{6.2}$	2–5	4/4	8, 8, 15, 15	—
	11–25	LLE46$^-$ pool 970	10$^{6.2}$	—	0/4	—	55
	11–25	SV40	10$^{6.2}$	5–19	4/4	19, 26, 26, 40	—
WHK *	12–17	LLE46 AG8	10$^{6.2}$	7	4/4	9, 9, 9, 12	—
	12–17	LLE46$^-$ pool 970	10$^{7.2}$	—	0/4	—	>63
WHK *	12–17	LLE46 AG8	10$^{6.2}$	7–12	4/4	12, 18, 20, 23	—
	12–17	LLE46$^-$ pool 970	10$^{7.2}$	—	0/4	—	>63
		LLE46 AG8 + SV40 anti-serum †	10$^{6.2}$	9–12	3/4	18, 18, 20, —	>63
		LLE46 AG8 + adeno 7 anti-serum †	10$^{6.2}$	—	0/4	—	>63

* These represent two different lots of WHK obtained from different sources.

† Virus was exposed to approximately 50 units of SV40 or Ad. 7 antiserum at room temperature for 45 min.; maintenance media contained antiserum.

TABLE II

SUMMARY OF EXPERIMENTS WITH ADENO 2, ADENO 3, AND ADENO 12-SV40 HYBRID VIRUSES

Tissue	Expt.	Inoculum	TCID$_{50}$/0.1 ml	Frequency of transformation no. tubes pos./ no. tubes inoc.	Day transformation first observed	Length of observation of neg. tubes
WHK	6–15	P.D. Ad.-3^{+} hybrid AG-3 Pool 1016	$10^{6.5}$	5/5	17	—
WHK	6–15	P.D. Ad. 3, Pool 5	$10^{6.7}$	—	—	83 days
WHK	7–7	Ad. 2^{++} R 2TILB + SV40 antiserum *	$10^{7.5}$	3/10	20	61 days
WHK	7–7	Ad. 2 Pool 1020	$10^{7.2}$	—	—	20 days
WHK	7–7		$10^{6.2}$	—	—	20 days
WHK	9–21	Ad. 2^{+} Pool 1017	$10^{5.5}$	5/5	14	—
WHK	9–21		$10^{4.5}$	3/5	38	76 days
WHK	9–21	Ad. 2 Pool 1020	$10^{5.2}$	—	—	76 days
WHK	9–21	Ad. 12^{+} Pool 1053	$10^{6.8}$	5/5	20	—
WHK	9–21		$10^{4.8}$	0/5	—	133 days
WHK	9–21	Ad. 12 Pool 3–453	$10^{6.5}$	1/3	109	133 days

* Virus was exposed to approximately 50 units of SV40 antiserum at room temperature for 45 min; maintenance media contained antiserum.

which may have been responsible for the transformation. Adeno 12^+ and adeno 12 both effected a neoplastic transformation; however, the transformation occurred both earlier and with a greater frequency with the adeno 12^+ preparation. In all these experiments, there was a variable, cytopathogenic effect (CPE) characteristic of the adenoviruses which, for the most part, regressed after involvement of 25 to 50% ($1+$ to $2+$) of the cell sheet. Adeno 2 at concentrations of $10^{7.2}$ and $10^{6.2}$ TCID$_{50}$ destroyed the entire sheet by the 6th to 9th day; at the highest dilution used ($10^{5.2}$ TCID$_{50}$) the $1+$ CPE which appeared by the 3rd to 4th day regressed after the 6th day following inoculation.

From the data on transformation of WHK with the hybrid viruses, we may conclude that the SV40 genome confers oncogenic potential to adenoviruses 7, 3, and 2 and that it enhances the oncogenic potential of adeno 12. The data with adeno 2 are especially interesting since this type is non-oncogenic. Although no transformation occurred with these strains of Ad. 7 and 3, other strains have been found to be oncogenic for hamsters.

Serologic Investigations

In Table III, the results of serologic studies carried out with the various hybrid transformed cell lines, some of the tumors induced by transplantation of these lines, and the sera of the tumor-bearing hamsters are given. SV40 T antigen, as determined by CF and FA tests, was present in all transformed lines and virtually every cell contained this antigen within its nucleus. Adeno T antigen was demonstrable only in the Ad. 12^+ transformed line. The inability to detect adeno T antigens, especially in adeno 7^+ and adeno 3^+ hybrid transformed cells with serum from hamsters bearing tumors induced with adenoviruses 7 and 3, respectively, may not be due to the absence of these antigens; tumors induced with the adeno 7^+ hybrid transformed cells had no detectable adeno 7 T antigen yet 4 of 28 hamsters developed antibody to this antigen. No adeno 2 T antigen was detected using the reagents enumerated in Table III.

SV40 T antigen was present in approximately half the tumors induced with Ad. 7^+ transformed cells and antibody to this antigen was present in the sera of these animals. Of the tumors induced with Ad. 2^{++} transformed cells which were not anticomplementary, all contained SV40 T antigen and some contained SV40 viral antigen and yielded infectious virus, presumably from the transcapsid particle which was not neutralized by SV40 antiserum.

Pathology of Tumors

Thirty tumors induced by 3 cell lines of Ad. 7^+ transformed WHK cells were examined histologically. These tumors consisted of a spectrum of morphologic variants. The majority of the tumors were composed of whorls and bundles of large spindle cells with indistinct cell margins and moderate amounts of pale acidophilic cytoplasm; these tumors resembled those produced in hamsters with SV40 virus (10–12) and some WHK lines transformed *in vitro* by SV40 virus

TABLE III

ANTIGEN IN CELL LINES AND TUMORS INDUCED WITH TRANSFORMED CELLS;
ANTIBODY IN SERA OF TUMOR-BEARING HAMSTERS.

			Ad. 7+	Ad. 3+	Ad. 2++	Ad. 2+	Ad. 12+
Antigen	Cell lines	CF					
		SV40 T	3/3 *	2/2	2/2	2/2	1/1
		SV40 V	0/3	0/2	0/2	0/2	0/1
		Adeno T	0/3 [1]	0/2 [2]	0/2 [3]	0/2 [3]	0/1 [4]
		Adeno V	0/3	0/2	0/2	0/2	0/1
		FA					
		SV40 T	3/3	2/2	2/2	2/2	1/1
		SV40 V	0/3	0/2	0/2	0/2	0/1
		Adeno T	0/3 [1]	0/2 [1]	0/2 [1]	0/2 [4]	1/1 [4]
	Tumors	CF					
		SV40 T	9/22 †		4/4		
		SV40 V	0/22		2/4		
		Adeno T	0/22 [5]		0/4		
		Adeno V	0/22		0/4		
	Antibody	CF					
		SV40 T	14/28 ‡		10/10		
		SV40 V	0/28		0/10		
		Adeno T	4/28		0/10 [6]		
		Adeno V	0/28		0/10		

[1] Reactions carried out with sera from hamsters bearing tumors induced with types 7 and 12.

[2] Reactions carried out with sera from hamsters bearing tumors induced with types 3, 7, and 12.

[3] No adenovirus 2 T antigen has yet been delineated; this test was carried out with Ad. 3, 7, and 12 hamster tumor sera.

[4] Adeno 12 hamster tumor serum used for these tests.

[5] Adeno 7 hamster tumor serum used for these tests.

[6] Reactions carried out with antigens prepared from hamster tumors induced with adenoviruses 3, 7, 12, and from antigens prepared from human embryonic kidney cultures infected with adenoviruses 2, 3, 7, 24 hours before harvest (cell pack antigen).

$$* \frac{\text{No cell lines positive}}{\text{No cell lines tested}} \quad † \frac{\text{No tumors positive}}{\text{No tumors tested}} \quad ‡ \frac{\text{No sera positive}}{\text{No sera tested}}$$

(13). The nuclei were large, pleomorphic, and vesicular, with margination of the nuclear chromatin and one to three prominent nucleoli. Mitoses, many of them bizarre, were frequent. Tumor giant cells typical of SV40 tumors were present in varying numbers; these cells had homogeneous acidophilic cytoplasm and several oval nuclei that often formed a syncytium.

A somewhat different anaplastic cell type was found in several of the tumors. The cells were arranged in broad sheets and were characterized by abundant pale acidophilic cytoplasm, round to polygonal cell margins, and typical SV40 type pleomorphic vesicular nuclei. This anaplastic component was quite prominent in some tumors, and did not have a counterpart cell type in tumors induced by SV40 transformed hamster kidney cells (13). In a few of these tumors there was a minor component which consisted of sarcomatous cells with vacuolated to clear cytoplasm.

Five of the 30 tumors had some evidence of epithelial differentiation, but without frank tubular or acinar formation. Irregular cords or pseudoacini were characteristically found adjacent to small areas of tumor necrosis. They were composed of spindle to cuboidal shaped cells with a small amount of brightly acidophilic cytoplasm and fairly uniform oval nuclei, and were easily distinguishable from the surrounding fibrosarcomatous component.

In addition to the features noted above, evidence of unequivocal adenocarcinoma was present in 5 tumors. Discrete tubules and acini were made up of cuboidal cells with regular, oval, hyperchromatic nuclei. These carcinosarcomas were indistinguishable from those induced by SV40 transformed hamster kidney cells (13).

The adenovirus 2^{++} transformed cells produced tumors with a histology similar to those produced by Ad. 7^+ but without evidence of frank epithelial differentiation. Two tumors had small areas of undifferentiated cells typical of neither SV40 nor adenovirus tumor, but there was no resemblance to the usual uniform small cell adenovirus tumor (14). One tumor was primarily composed of columns of poorly differentiated epithelial-like cells that blended with the adjacent stroma. This cell type was not noted in any of the other tumors.

Discussion

From the data presented, it can be concluded that the various adeno-SV40 hybrid viruses transformed WHK tissue culture cells. Since the strains of adenoviruses 2, 3, and 7 used in these experiments did not effect transformation, the aquisition of a portion of the SV40 genome, although resulting in a defective particle with respect to infectivity, has apparently conferred an oncogenic potential to these viruses. Studies are in progress to determine the kinetics of transformation with the hybrid populations. If transformation follows one-hit kinetics the hybrid particles could be unequivocally identified as the transforming particle; infectivity in AGMK cultures follows two-hit kinetics indicating that the hybrid and complete adenovirions are needed for an infectious center (3, 4).

The acquisition of a part of the SV40 genome by adenovirus 12 resulted in an enhanced transforming potential. A similar type of enhancement has been found *in vivo;* tumors developed with much shorter latent periods and in a larger number of hamsters inoculated with both adenovirus 12 and SV40 than with either virus alone (15). Enhancement of adenovirus infectivity in AGMK cultures by SV40 virus has also been described (18).

The type of transformation resulting from adeno 2, 3, and 7 hybrids was characteristic of that produced by SV40 (1). Moreover, the tumors induced with adeno 7 hybrid transformed cells were similar to those described with SV40 transformed WHK cultures (13 and unpublished observations). Tumors from one of these 3 lines contained tubules; such epithelial differentiation has been observed with SV40 transformed WHK cells which contained the complete SV40 genome. The occurrence of similar tubules with transplanted cells transformed by hybrid virus containing only a portion of the SV40 genome indicates that the whole genome is not necessary for the presumed disturbance of differentiation (13, 19). In addition, these tumors contained some components reminiscent of adenovirus tumors. Thus, they differed from the tumors examined by Rabson *et al.* which were composed of a small round cell similar to that seen in primary hamster tumors induced with adenoviruses (16). The transformations by adeno 12 and 12$^+$ were similar and are in keeping with an enhancement of a malignant potential already present. Tumors induced by these transformed lines have not been examined.

Since a part of the SV40 genome presumably integrated with the adenovirus DNA has conferred an oncogenic potential to the adeno 2, 3, and 7 viruses resulting in an SV40 type of morphological transformation, it was hypothesized that the SV40 genome was responsible (1). While this may be true for the adeno 7$^+$ hybrid, a possible criticism of this hypothesis as regards the adeno 2$^+$ and 12$^+$ hybrids stems from recent reports demonstrating linkage between the SV40 and adeno 7 genomes (19). Thus, a portion of the adeno 7 DNA was transferred with the SV40 genome to both the adeno 2 and adeno 12 viruses when these hybrids were produced (Rowe and Pugh, personal communication). It is conceivable that this portion of the adeno 7 DNA contributed to the malignant potential of the hybrids and this is currently under investigation.

Studies are also in progress to determine whether the SV40 specific transplantation rejection antigen is present in these transformed cells. Rapp and Melnick have shown transplantation resistance for SV40 tumor cells following immunization with the Ad. 7$^+$ hybrid preparation; thus, it would appear that the hybrid virus is capable of inducing SV40 transplantation antigen. If this finding is confirmed, it would mean that enough SV40 genetic information is present to code for at least 2 proteins since the T and transplantation antigens are distinct (17).

In conclusion, *in vitro* transformation by the adeno-SV40 hybrid viruses offers a unique opportunity to study the effects of the various nucleic acid components present as regards target cell type, morphology of resulting transformed cells and the tumors induced by transplantation of these cells, enhancement of oncogenicity, and antigen formation—in particular, possible T antigen formation by non-oncogenic viruses. The utilization of other markers of SV40 DNA, such as the transplantation resistance antigen in transformed cells, offers an opportunity to determine the amount of SV40 genetic material present. In addition, hybrid transformed cells may offer an opportunity to study any relationship between the transplantation rejection antigen(s) induced by the various types of oncogenic adenoviruses.

References

1. BLACK, P. H., and TODARO, G. J.: *In vitro* transformation of hamster and human cells with the adeno 7-SV40 hybrid virus. Proc. Nat. Acad. Sci. USA **54**, 374 (1965).
2. HUEBNER, R. J., CHANOCK, R. M., RUBIN, B., and CASEY, M. J.: Induction by adenovirus type 7 of tumors in hamsters having the antigenic characteristics of SV40 virus. Proc. Nat. Acad. Sci. USA **52**, 1333 (1964).

3. ROWE, W. P., BAUM, S. G., PUGH, W. E., and HOGGAN, M. D.: Studies of adenovirus-
SV40 hybrid viruses. I. Assay system and further evidence for hybridization. J. Exp.
Med. 122, 943 (1965).

4. ROWE, W. P., and BAUM, S. G.: Studies of adenovirus-SV40 hybrid viruses. II. Defectiveness of the hybrid particles. J. Exp. Med. 122, 955 (1965).

5. BOEYÉ, A., MELNICK, J. L., and RAPP, F.: Adenovirus-SV40 "hybrids": Plaque purification into lines in which the determinant of the SV40 tumor antigen is lost or retained.
Virology 26, 511 (1965).

6. ROWE, W. P., and BAUM, S. G.: Evidence for a possible genetic hybrid between adenovirus type 7 and SV40 viruses. Proc. Nat. Acad. Sci. USA 52, 1340 (1964).

7. RAPP, F., MELNICK, J. L., BUTEL, J. S., and KITAHARA, T.: The incorporation of SV40
genetic material into adenovirus 7 as measured by intranuclear synthesis of SV40 tumor
antigen. Proc. Nat. Acad. Sci. USA 52, 1348 (1964).

8. ROWE, W. P.: Studies of adenovirus-SV40 hybrid viruses. III. Transfer of SV40 gene
between adenovirus types. Proc. Nat. Acad. Sci. USA 54, 711 (1965).

9. RAPP, F., BUTEL, J. S., and MELNICK, J. L.: SV40-adenovirus "hybrid" populations:
Transfer of SV40 determinants from one type of adenovirus to another. Proc. Nat. Acad.
Sci. USA 54, 717 (1965).

10. EDDY, B. E., BORMAN, G. S., BERKELEY, W. H., and YOUNG, R. D.: Tumors induced in
hamsters by injection of Rhesus monkey kidney cell extracts. Proc. Soc. Exp. Biol. Med.
107, 191 (1961).

11. GIRARDI, A. J., SWEET, B. H., SLOTNICK, V. B., and HILLEMAN, M. R.: Development
of tumors in hamsters inoculated in the neonatal period with vacuolating virus SV40.
Proc. Soc. Exp. Biol. Med. 109, 649 (1962).

12. BLACK, P. H., and ROWE, W. P.: Viral studies of SV40 tumorigenesis in hamsters. J. Nat.
Cancer Inst. 32, 253 (1964).

13. BLACK, P. H., ROWE, W. P., and COOPER, H. L.: An analysis of SV40-induced transformation of hamster kidney tissue in vitro. II. Studies of three clones derived from a
continuous line of transformed cells. Proc. Nat. Acad. Sci. USA 50, 847 (1963).

14. TRENTIN, J. J., YABE, Y., and TAYLOR, G.: The quest for human cancer viruses. Science
137, 835 (1962).

15. SCHELL, K., LANE, W. T., CASEY, M. J., and HUEBNER, R. J.: Potentiation of
oncogenicity of adenovirus type 12 grown in African Green Monkey kidney cell cultures
preinfected with SV40 virus: Persistence of both T antigens in the tumors and evidence
for possible hybridization. Proc. Nat. Acad. Sci. USA 55, 81 (1966).

16. RABSON, A. S., MALMGREN, R. A., and KIRSCHSTEIN, R. L.: Induction of neoplasia
in vitro in hamster kidney tissue by adenovirus 7-SV40 "hybrid" (strain LLE 46). Proc.
Soc. Exp. Biol. Med. 121, 486 (1966).

17. TEVETHIA, S. S., KATZ, M., and RAPP, F.: New surface antigen in cells transformed by
simian papovavirus SV40. Proc. Soc. Exp. Biol. Med. 119, 896 (1965).

18. RABSON, A. S., O'CONOR, G. T., BEREZESKY, I. K., and PAUL, F. J.: Enhancement of
adenovirus growth in African Green Monkey kidney cell cultures by SV40. Proc. Soc.
Exp. Biol. Med. 116, 187 (1964).

19. BLACK, P. H.: J. Nat. Cancer Inst., in press (1966).

Cytogenetic Alterations During Malignant Transformation

By

PAUL S. MOORHEAD AND DAVID WEINSTEIN

Wistar Institute of Anatomy and Biology
Philadelphia, Pennsylvania

Interest in the degree of chromosomal stability of normal cells *in vitro* and in the same cells under conditions leading to "malignant transformation" obviously relates to various theories of cancer which can be assembled under the term "somatic mutation." Only in the past 7 or 8 years have techniques been available for adequate studies of mammalian chromosomes. The comments of Schultz (18) are appropriate in this regard: "It is a truism by now that the change from the normal to the neoplastic cell must involve a change in cellular heredity. However, this statement is now so general as to be meaningless; it demands a rephrasing in concrete terms and a more definite conceptual analysis." Schultz pointed out that alternative hypotheses often consist of nothing more than restatements "specified in terms of a particular phenotype" which also applies to the viral theory of tumorigenesis. The virus may act only at one point during infection and its subsequent loss would be immaterial since it functions as an initiator, an agent of gene or structural mutation. Viral DNA may possibly become incorporated into the mammalian host cell genome in the manner of lysogenic systems but again this can be classified as a *special* case of a somatic cell mutation.

Formalized genetic mechanisms can plausibly account for both initiation and further progression to the fully malignant state, which thus constitutes an embarrassment of riches since we have so many possible mechanisms to distinguish within the framework of "genetic change." Aside from gene mutation, polyploidy and aneuploidy can disturb the gene balance within the cells. Shifts in gene balance that would be lethal at the organ or organism level may be tolerated and perpetuated at the cell level. Chromosome breakage and reunion can lead to translocations, deletions, inversions, etc., which not only affect gene balance but also open the possibilities of position effects. Transposition of controlling genes, i.e., genes concerned with differentiation processes, are of considerable conjectural interest in this respect. This interlocking of differentiation and genetic change may yet be distinguished from hypotheses *limited* to differentiation considered as a true modulation of gene expression.

The state of the present discussion centers upon which of various genetic interpretations is favored by evidence from studies of chromosomal events. Furthermore, does evidence exist which implies irreversible shifts of cell differentiative machinery entirely apart from structural rearrangement and mutation?

104

In our own work we have tended to equate "transformation" of mammalian cells from the "normal" cell to its malignant descendant. However, it has become obvious that there is no single pattern or sequence of steps which can uniformly be applied to the various mammalian cells under study. Rather than restrict our definition to the whole series of observable or measurable events, even though these may be reproducible for any one system, "transformation" can be defined as the acquisition of any one of several "unit characters" or elements which is thereafter heritable.

Fould's concept (5) of tumor progression in terms of "unit characters" is the basis for our concept that *in vitro* transformation can serve as an experimental model of neoplasia. The unit characters are, in a sense, crude phenotypes and may involve any of the following: a) *cell morphology* such as cell size, cell shape, nuclear pleiomorphism, b) *culture morphology or cell interactions*, such as loss of contact inhibition of migration, c) *antigenic changes*, d) *chromosome changes* including breakage, polyploidy, aneuploidy, e) *growth* as exhibited by stimulation of mitosis, rate of mitosis, plating efficiency, release of mitosis from normal inhibition, and f) *malignancy* as shown by transplantability. We are concerned with the determination of *in vitro* unit characters which may be sufficient or necessary for the commitment of a cell to a course which culminates in malignancy. There is little agreement as to which characters may be extrapolated to *in vivo* malignancy or even as to which are coupled. Much of this doubt stems from the problem of assessing *in vivo* malignancy itself, an obvious limitation for human cell transformations.

Chromosomal studies of *in vitro* transformation of mammalian cells can be said to have begun with the work of Levan and Biesele (11) in which they outlined the general features of spontaneous transformation in the mouse cell. These were: the progressive loss of euploidy and the emergence of abnormal cell lines which were autonomous *in vitro* and most of which were malignant upon implantation into appropriate hosts. Subsequent work by Rothfels, et al. (16), and by Todaro and Green (22) confirmed and supported this general expectation for mouse cells in culture, i.e., the acquisition of aneuploidy and enhanced growth *in vitro* and *in vivo*. In mouse cell cultures the occurrence of some chromosome abnormalities clearly precedes the changes in growth ability although a general change of the culture's ploidy is not complete until the shift in growth ability has already taken place (22). The use of cultures of mouse cells places upon the experiment the additional burden of resolving *initiation* of a process from *acceleration* of that process. In any case, these studies clearly demonstrate that the resulting general heteroploidy, or more properly mixoploidy, is not the direct basis for the enhancement of growth observed in spontaneous transformation.

In fact, in one of Rothfels' established mouse lines which was heteroploid or mixoploid there was no apparent increased *rate* of growth. Defendi has emphasized a number of exceptions to the general experience regarding the lack of correlation between the various elements of the complete transformation process (2). Other instances have been mentioned within this symposium. The most interesting examples of this lack of coupling of the elements of general transformation are the BHK/21 hamster cell line and the 3T3 mouse cell line (21, 23). The BHK/21 line possesses many of the unit characters regarded as parameters of transformation in other systems. Although this cell shows a higher plating efficiency than does the primary hamster cell and approximately 10^6 cells may induce a tumor, it retains quite clearly its fusiform cell shape and parallel orientation of cell growth. After some 200 *in vitro* generations this morphological character is still preserved even though approximately 5% of the cells are recognizably aneuploid or quasi-diploid and as few as 10 cells may induce tumors on implantation (4). Thus, it can be regarded as an unstable cell which possesses only a part of the full set of transformation elements encountered in other cell-virus systems. The 3T3 mouse cell line is completely heteroploid, has a high *in vitro* plating efficiency and yet it retains its "normal" capacity to show contact inhibition of migration and contact inhibition of cell division. Each of these continuous lines provides an excellent assay system for studies of "transformation," but, it must be kept in mind that each has acquired certain of the "unit characters" which define transformation in other systems, and in some instances even in hamster and mouse systems.

The use of viruses has provided the only agent permitting controlled study of transformations during the entire process. I will discuss in some detail the systems with which I am most familiar, the SV40-human cell system and the polyoma-hamster cell system. Both involve nearly all of the major steps of change in cell phenotype associated with transformation. Defendi and Lehman (3) have kindly permitted me to discuss their results with the polyoma-hamster cell system, a virus-cell system which is as complete as the SV40-human cell system but which has the advantage of permitting adequate *in vivo* tests of malignancy with inbred Syrian hamster lines.

SV40 and Human Diploid Fibroblast-like Cells

Mass cultures of human fibroblast-like cells which were infected and subcultured routinely thereafter became "transformed," in the ordinary sense of gross morphological changes and more efficient tissue culture growth (10, 19). Transformation is invariably accompanied by extensive chromosomal abnormalities. Embryonic or adult human fibroblast-like cell cultures have the advantage of a zero rate of spontaneous transformation *in vitro*, under the conditions of culture ordinarily used, and also have a very low background rate of chromosomal breaks or other aberrations during most of their characteristically limited *in vitro* lifespan. Continued growth routinely fails to occur in our laboratory for the WI-38 cell strain at 42 to 52 average cell doublings. The percentage of spontaneous chromatid and chromosome breaks is 4.5% but during the last 5 to 10 subcultures it increases to about 10 or 25% and derivative aberrations such as dicentrics and exchanges are also characteristic of this final degenerative stage of *in vitro* culture (17).

Infection with SV40 virus during this latter period of declining growth ability leads to a much more rapid transformation, in terms of growth and morphology (1–2 weeks versus 8–12 weeks). This correlation between exhaustion of growth potential and a shorter elapsed time between infection and transformation was noted independently by Todaro, Wolman and Green (24) and by Jensen and coworkers (9).

Extensive chromatid and chromosome breakage is induced and among the earliest observed results is the appearance of a fraction of cells which have lost one or more acrocentrics (28, 13); however, this phenomenon is not always observed and is most likely due to a higher survival value of such cells among the various genetically mutilated cells. Also observed are polyploidy which is a consistent observation, and the attenuation of one type of secondary constriction. Derivative types of abnormal, unstable chromosomes, such as dicentrics, occur in high frequency.

In an attempt to determine the relative sequence of the various cytologically observable changes serial studies were done with WI-38 following infection of a mass culture with SV40 (13). The usual delay or "silent" period of some weeks was followed by: a) coincident increases in chromatid breaks, in tetraploidy, and in cell division, b) the expected occurrence of high frequencies of the derived types of chromosome abnormalities, and c) apparent escape from "senescent" changes, that is, establishment as a permanent line.

Similarly, in an experiment involving infection of an "older" culture (Passage 30) chromatid breaks and tetraploidy increased; derivative abnormal chromosomes were coincident but in this instance preceded the earliest detection of morphological changes, principally based upon crude observations of widespread foci of high mitotic rate.

In the assay systems first developed for viral transformation in mouse cell cultures the key event seemed to have occurred shortly after infection and morphology changes were evident in a very short time (25, 26). Similarly, as mentioned above, a short elapsed time between infection and transformation in human fibroblast-like cells using SV40, if aged cultures were infected. Wolman, Hirschorn and Green (27) found dicentric chromosomes only 5 days after infection, although spontaneous background aberrations in such "older" cultures may have contributed to this.

The long silent period after infection of mass cultures during the rapid growth period (Phase II) could be due to a) a low efficiency of adsorption and/or penetration, b) a low capacity of response by all cells or c) differential capacity of response by only a fraction of the cells (target cells). Also involved are questions of inhibition of transformed cells by the Phase II type population (19) and of the time necessary for emergence of those cells committed to the process of transformation.

In the past year, a third long-term study of SV40 transformation of human cells, WI-38, was done with greater attention to virus-cell dynamics (7). The results of this study can be summarized as follows: "The initial infection, adsorption, and synthesis of new virus occurred in the absence of any *discernable* chromosome changes or growth alteration. The lack of early chromosome changes following infection of Phase II cultures had been assumed from previous studies since little or no cytopathic effect occurred and there was always an elapsed time of some weeks between infection and transformation. Chromosomal analysis at 6, 13, 20 and 24 hours and every few days thereafter following exposure to SV40 virus produced no evidence of any chromosomal damage. Therefore, it seems reasonable to exclude the possibility that early effects upon the chromosomes had escaped detection."

"The acquisition of new growth properties occurred in two distinct steps. The first of these was detectable 48 days after infection as the loss of the restriction upon cell division that is associated with, and presumably imposed by, the confluent monolayer in such fibroblast cultures."

"The loss of inhibition of division occurs two days before the first observation of any chromosomal abnormalities and may possibly reflect only a greater sensitivity for determination of loss of contact inhibition of division. The relatively short time interval separating some of the elements defining transformation makes it doubtful whether firm conclusions can be based upon the particular chronology observed here. Obviously, *sensitivities for detection* of various parameters may differ. Further, the resolution of points in the time scale depends upon the frequency of sampling; in this case, approximately four days separated each sample. Whether chromosomal damage exists in those cells first observed to be dividing at the time of loss of contact inhibition of division is now under study. Chromosomal abnormalities appear two days after the initial sign of transformation and there is also an increase in frequency of tetraploids. Both of these trends persist, and their initial appearances precede any discernable alteration of cell morphology. The populations chromosome number can be termed heteroploid on day 57, whereas morphologic changes could not be ascertained from coverglass preparation until day 70. This was true even though detection of significant changes in chromosome number constitutes a rather insensitive measure of transformation, since the cytologic techniques used do produce a background of artificially subdiploid counts. Possible disturbances of the spindle are indicated by the increase in tetraploidy observed in the sample taken only two days after loss of inhibition of division. However, the presence of diplochromosome metaphase sets (endoreduplication) among the tetraploids may mean that a disturbance of DNA synthesis rather than of the spindle was more important among the consequences of the presence of the virus."

"Although available data seems to substantiate the fact that the *primary* change in the ability of cells to replicate is due to a failure to recognize a control signal, the nature of the *second* growth step is unclear. An increase in the mitotic index did occur just prior to the second increment in growth capacity; however, increase in observed percentage of mitotic cells may conceivably reflect any one, or combination of events: (a) simple decrease in generation time; (b) stimulation to division of cells previously not participating; (c) delay in metaphase stage due to pathologic interference, as observed in some radiation studies. The second increment in growth ability attained by the infected line coincided in time with the first alteration in cellular morphology, at about 70 days; and it cannot be excluded that both properties were conferred upon the evolving population by changes in the cell genome arising from the extensive alteration of chromosome number and form then present. Whether these new properties exhibited by the population derive from deletions and rearrangement of existing genetic information or whether the gross rearrangements are merely indicators of even more extensive point mutation events (14) is a question for future investigation."

"Since a transient increase in *chromatid* breakage had been observed in the two previous experiments of this type (13), it is not understood why a similar occurrence was not observed in the present study."

In experiments involving observations even within 1 or 2 generation periods following infection reproducibility in respect to chromosome damage is not always attained. Vogt and Dulbecco (26) observed increases in anaphase aberrations (rod and dot deletions and bridges) from a level of about 10% in the controls to 50–70% very soon after infection of Syrian hamster cells with polyoma virus. No such marked increase in breakage scored at metaphase of similarly infected cells was observed by Defendi and Lehman (4). The scoring of breaks as precursors of the more obvious rearrangements has been observed to fluctuate in other experimental situations also. Measles-induced breakage *in vivo* in humans has been observed (15) and confirmed (1) but a number of other competent workers have not observed any elevation of breaks (8). This lack of clear reproducibility may relate to factors which affect the restitution of breaks. It is known from radiation studies that for each cytologically visible break perhaps of the order of 100 breaks may have healed.

The polyoma-Syrian hamster cell system demonstrates conclusively that the continued production of chromosomal anomalies does not depend upon the continued presence of infectious virus. This is true also for the SV40-human cell system and others, but the amount of complete polyoma virus recoverable decreases to nil in a very short time in the face of extensive chromosomal rearrangement.

The continued instability and production of new chromosomal anomalies is observed in nearly all of the spontaneously established human cell lines and also in the permanent human cell lines arising from SV40 transformation; however, this generalization needs qualification in one respect. The *degree* of involvement and the range of anomalies is much greater in virus-transformed cells in the weeks or months following initial transformation than in the permanent cell line which finally results. After initial loss of inhibition of mitosis and initial karyological change there is usually a period of quite bizarre chromosomal "invention" before the culture fails. This delayed degenerative period is nearly always observed and has been termed "crisis" in order to distinguish it from a true lytic phase, even though virus is still being shed (6, 20). During "crisis" the mitotic rate observed among cells which remain in the monolayer is very high (20), about 8% in our material, even though the total cell number is decreasing rapidly due to cell detachment.

From the point of view of the chromosomal events, this degenerative event might be attributed to the elimination of many lethal combinations produced by the extensive restructuring throughout the weeks preceding crisis. Tetraploidization of the population would tend to have the opposite effect since duplication of gene dose may buffer the effects. However, elimination of lethals seems too glib an explanation in view of a recent finding: the occurrence of crisis is not related to the time of infection but to the time of Phase III degeneration in the *uninfected* sister cultures (Girardi, pers. communication). This observation is interesting since it suggests that the cumulative loss of growth capacity *in vitro*, characterizing the original human cells, is delayed by viral transformation for weeks or even months but does express itself even after marked reshuffling of the chromosomal material. Therefore, the *in vitro* "ageing effect" which is correlated with a more effective transformation by SV40 virus, seems to be independent of the genetic information of the cell—within wide limits. From initial chromosomal changes to the period of crisis, 90 days in the experiment described, the transformed population completed a minimum of 30 average cell doublings. Selective pressures within this period and those involved in recovery from crisis would be extreme. Usually only a few islets of surviving colonies finally emerge in a culture vessel. This extensive evolution and selection *in vitro* may correspond to *in vivo* progression. A general correlation between degree of *in vitro* aneuploidy and malignancy was observed in the polyoma-hamster cell system (3). Also, the grade of malignancy of rabbit-papilloma-virus tumors could be correlated with degree of chromosome change in an *in vivo* study (12).

Although aneuploidy and visible chromosome damage are not *invariably* coupled during the establishment of cell lines *in vitro* nor with all tumors, the very high correlation of these remains unexplained. Aside from oncogenic progression the hypothesis that a minimum number of cryptostructural lesions are necessary for the inactivation of some key genetic factor cannot yet be excluded.

References

1. AULA, P.: Virus-associated chromosome breakage. A cytogenetic study of chickenpox, measles and mumps patients and of cell cultures infected with measles virus. Ann. Acad. Sci. Fenn. Series A, IV Biologica #89:1–75 (1965).

2. DEFENDI, V.: Transformation *in vitro* of mammalian cells by polyoma and simian 40 viruses. Prog. in Exp. Res. 8, *in press*.

3. DEFENDI, V., and LEHMAN, J. M.: Transformation of hamster embryo cells *in vitro* by polyoma virus: morphological, karyological, immunological and transplantation characteristics. J. Cell. & Compar. Physiol. Dec. (1965).

4. DEFENDI, V., LEHMAN, J., and Kraemer, P.: "Morphologically normal" hamster cells with malignant properties. Virology 19:592–598 (1963).

5. FOULDS, L.: The experimental study of tumor progression: a review. Cancer Res. 14:327–339 (1954).

6. GIRARDI, A. J., JENSEN, F. C., and KOPROWSKI, H.: SV-40-induced transformation of human diploid cells: Crisis and recovery. J. Cell. & Compar. Physiol. 65:69–83 (1965).

7. GIRARDI, A. J., WEINSTEIN, D., and MOORHEAD, P. S.: SV40 transformation of human diploid cells: Parallel studies of viral and karyological parameters. Annales Medicinae Experimentalis et Biologiae Fenniae, *in press* (1966).

8. HARNDEN, D. G.: Cytogenetic studies on patients with virus infections and subjects vaccinated against yellow fever. Amer. J. Human Genetics 16:204–213 (1964).

9. JENSEN, F., KOPROWSKI, H., and PONTEN, J. A.: Rapid transformation of human fibroblast cultures by simian virus 40. Proc. Natl. Acad. Sc. 50:343–348 (1963).

10. KOPROWSKI, H., PONTEN, J. A., JENSEN, F., RAVDIN, R. G., MOORHEAD, P. S., and SAKSELA, E.: Transformation of cultures of human tissue infected with simian virus SV40. J. Cell. and Compar. Physiol. 59:281–292 (1962).

11. LEVAN, A., and BIESELE, J. J.: Role of chromosomes in cancerogenesis, as studied in serial tissue culture of mammalian cells. Ann. N. Y. Acad. Sci. 71:1022–1053 (1958).

12. MCMICHAEL, H., WAGNER, J. E., NOWELL, P. C., and HUNGERFORD, D. A.: Chromosome studies of virus-induced rabbit papillomas and derived primary carcinomas. J. Natl. Cancer Inst. 31:1197–1214 (1963).

13. MOORHEAD, P. S., and SAKSELA, E.: The sequence of chromosome aberrations during SV40 transformation of a human diploid cell strain. Hereditas 52:271–284 (1965).

14. NICHOLS, W. W.: Relationships of viruses, chromosomes, and carcinogenesis. Hereditas 50:53–80 (1963).

15. NICHOLS, W. W., LEVAN, A., HALL, B., and OSTERGREM, G.: Measles-associated chromosome breakage. Hereditas 48:367–370 (1962).

16. ROTHFELS, K. H., KUPELWIESER, E. B., and PARKER, R. C.: Effects of X-irradiated feeder layers on mitotic activity and development of aneuploidy in mouse-embryo cells in vitro. Canad. Cancer Conf. 5:191–223 (1963).

17. SAKSELA, E., and MOORHEAD, P. S.: Aneuploidy in the degenerative phase of serial cultivation of human cell strains. Proc. Nat. Acad. Sc. 50:390–395 (1963).

18. SCHULTZ, J.: Malignancy and the genetics of the somatic cell. Ann. N. Y. Acad. Sci. 71:994–1008 (1958).

19. SHEIN, H. M., and ENDERS, J. F.: Multiplication and cytopathogenicity of simian vacuolating virus 40 in cultures of human tissues. Proc. Soc. Exp. Biol. & Med. 109:495–500 (1962).

20. SHEIN, H. M., ENDERS, J. F., PALMER, L., and GROGAN, E.: Further studies on SV40-induced transformation in human renal cell cultures. I. Eventual failure of subcultivation despite a continuing high rate of cell division. Proc. Soc. Exp. Biol. & Med. 115:618–621 (1964).

21. STOKER, M., and MACPHERSON, I.: The Syrian hamster fibroblast cell line BHK/21 and its derivatives. Nature 203:1355–1357 (1964).

22. TODARO, G. J., and GREEN, H.: Quantitative studies of the growth of mouse embryo cells in culture and their development into established lines. J. Cell. Biol. 17:299–313 (1963).

23. TODARO, G. J., GREEN, H., and GOLDBERG, B. D.: Transformation of properties of an established cell line by SV40 and polyoma virus. Proc. Natl. Acad. Sc. 51:66–73 (1964).

24. TODARO, G. J., WOLMAN, S. R., and GREEN, H.: Rapid transformation of human fibroblasts with low growth potential into established cell lines by SV40. J. Cell. & Compar. Physiol. 62:257–265 (1963).

25. VOGT, M., and DULBECCO, R.: Virus-cell interactions with a tumor-producing virus. Proc. Natl. Acad. Sc. 46:365–370 (1960).

26. VOGT, M., and DULBECCO, R.: Steps in the neoplastic transformation of hamster embryo cells by polyoma virus. Proc. Natl. Acad. Sc. 49:171–179 (1963).

27. WOLMAN, S. R., HIRSCHORN, K., and TODARO, G. J.: Early chromosomal changes in SV40-infected human fibroblast cultures. Cytogenetics 3:45–61 (1964).

28. YERGANIAN, G., SHEIN, H. M., and ENDERS, J. F.: Chromosomal disturbances observed in human fetal renal cells transformed in vitro by simian virus 40 and carried in culture. Cytogenetics 1:314–325 (1962).

Relative Growth and Oncogenic Potential of Normal, Malignant, and Virus-Transformed Euploid Cells of Dwarf Species of Hamster[1]

(Discussion to Dr. Paul Moorhead's Presentation on Cytogenetic Alterations During Malignant Progression)

By

GEORGE YERGANIAN

Laboratories of Cytogenetics, The Children's Cancer Research Foundation
Harvard University Medical School
Boston, Massachusetts

Introduction

Numerical and structural alterations of the karyotype are secondary phenomena featured among many, if not all, continuously propagated normal, malignant, and virus-transformed cell lines. Dr. Moorhead has described the varied events during *in vitro* progression of human and Syrian hamster virus-transformed cells. His efforts to associate cytogenetic alterations with increased oncogenicity must, however, be restricted to the species employed for reasons I will describe below. The degree of aneuploidy need not be extensive, in order for experimental cells to proliferate more rapidly, or to attain the status of a permanent cell line. I sense that Dr. Moorhead will agree to the statement that a subtle, stable pseudodiploid condition is more conducive to the development of a cell or tumor line than many complex heteroploid conditions (Yerganian, *et al.*, 1960; Moorhead, 1965).

The degree to which the chromosome complement of virus-transformed cells may alter is strongly influenced by unknown species-specific factors. In addition, the relative ease (or difficulty) in establishing long-term cultures from normal derivatives also bears strongly on this subject. Using these as indices, the human and hamster fibroblast cell systems discussed at this meeting may be grouped as follows: (a) control cultures remain euploid while virus-transformed cells undergo chromosomal alterations shortly thereafter, as described by Dr. Moorhead for human cell strains; (b) control cultures generally fail to continue and the majority of transformed cultures may display karyotypic changes, as reviewed by Dr. Moorhead for Syrian hamster cells; (c) control cultures of the Armenian hamster fail to

---

[1] Supported by grants from the Damon Runyon Memorial Fund (293), U.S.P.H.S. CA-08378 and C-6516, National Cancer Institute, and National Science Foundation (GB4558).

continue beyond the third passage (Yerganian and Papoyan, 1965), while transformed cells retain euploidy for 50 or more passages (Yerganian, et al., unpublished), and (d) controls and transformed cells (isolated as focal proliferations) of the Chinese hamster remain euploid for 20 to 125 passages (Yerganian, 1963a, b; Yerganian, et al., 1964, in press).

The latter cell systems, derived from dwarf species of hamsters, contrast sharply with Dr. Moorhead's description for the rapid appearance of cytogenetic alterations in virus-transformed human and Syrian hamster fibroblast cells. Following the exposure of primary and secondary cultures to a given tumor virus, the initial events leading to morphological and malignant transformation undoubtedly occur while cells are still euploid (Sachs, 1965). It therefore seems logical to assume that any attempt to clarify these events may best be done by employing a cell system which transforms readily and remains euploid for many cell generations. Perhaps the most crucial test to evaluate the potential of transformed euploid cells would be (a) to determine baseline levels of viral and cellular antigens, and (b) when variants do appear, clone for stable aneuploid sublines and repeat the tests. In the event stable aneuploids (single and double transformants) are found to have different levels of antigen production, one may attempt to correlate the relative increase or decrease in antigen(s) with the presence or absence of specific chromosome types. On the other hand, if euploids and aneuploid sublines have the same levels of antigen, further studies on chromosomes of transformed cells would be of little value, other than for mapping localized breakage and "gap" formation along chromosomes of cells exposed to viruses (refs. in Nichol's discussion, pp. 17–18).

Observations

Dwarf species of hamsters, namely, the Chinese hamster, *Cricetulus griseus* (2n = 22), and the Armenian hamster, *Cricetulus migratorius* (2n = 22), are excellent sources for normal, malignant and virus-transformed euploid cells. Idiograms of male karyotypes are presented in Fig. 1. This discussion is concerned primarily with data compiled for the Chinese hamster and reference to cell types of the Armenian hamster will be brief since this species is relatively new in our hands.

Control cultures. Normal fibroblast derivatives from adult lung and whole embryo minces proliferate readily as euploids for some 20–40 passages, depending upon culturing conditions and organ of derivation (Ford and Yerganian, 1958; Yerganian and Leonard, 1961; Hsu and Zenzies, 1964; Zakharov, et al., 1964). As long as euploidy persists (85+%) plating efficiencies rarely exceed the 10% level when seeding 1000–2000 cells per 60 mm. Petri dish. Contact inhibition and lack of oncogenicity also prevail during this period. When numerical alteration is noted, it may vary from subtle pseudodiploids involving a small metacentric and one of the subtelocentrics, and proceed to higher aneuploids, with or without subsequent deletion of the whole or either arm of the genetically-repressed X_2 or Y, in addition to aberrancies involving other autosomal types.

The time required for aneuploid sublines to replace the parental diploid population varies considerably and is influenced by the organ of origin (Stone, unpublished;

CRICETULUS GRISEUS

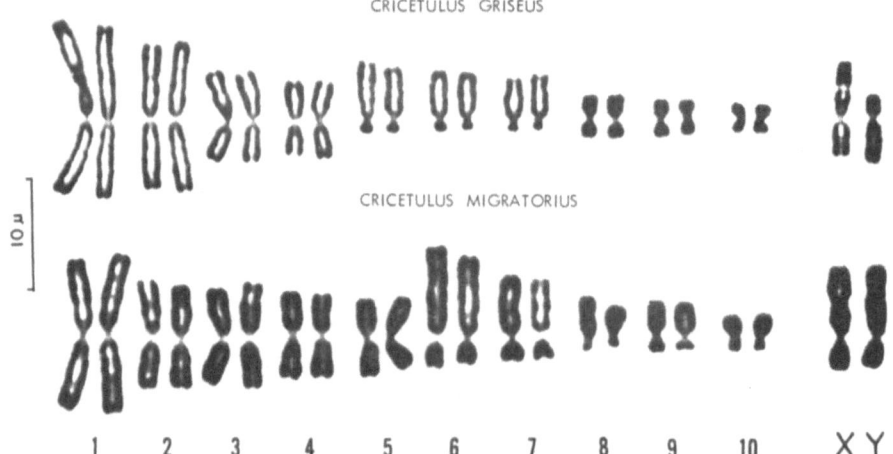

CRICETULUS MIGRATORIUS

10 μ

1   2   3   4   5   6   7   8   9   10   X Y

Fig. 1. Idiograms of normal male karyotype of Chinese (*Cricetulus griseus*) and Armenian (*Cricetulus migratorius*) hamsters. Alignment of Y chromosome in former species arranged to conform with pairing affinity of X and Y chromosomes during meiotic prophase (Yerganian and Lavappa, unpublished).

Yerganian, unpublished). Long-term retention of euploidy (up to one year), or an early onset of aneuploidy (by the 10th–20th passages) is strongly dependent upon the cell-culturing methods employed. Invariably, a single mutant (bearing a marker chromosome) rises to challenge the parental euploid stem cell (this contrasts to the array of heteroploids described by Dr. Moorhead where selection for a new stem cell karyotype is reduced because of perpetual nondisjunction). Both the parent euploid and challenging eneuploid persist in Chinese hamster cultures for as many as 10 passages, during which time both cell types can be cloned and continued thereafter as stable sublines for many months.

Aneuploidy is reflected by a number of physiological changes. For example, while euploid, 12E cells had a plating efficiency of 10%. By the 24th passage, trisomy for chromosome 8 was recorded in 84% of the metaphases and plating efficiency rose to 20%. By the 35th passage, a very small "minute" marker found its way into the karyotype. It consisted of nothing more than a centromere with the barest amount of visible chromatin attached to both "arms." Now, the plating efficiency increased to 50%. In time, variants can be selected which have plating efficiencies of 80+%, as noted among the more aged cell lines of this species employed for radiation-cell survival studies by Elkind, Puck, Sinclair, *et al.* Improved plating efficiency of aneuploids is considered to reflect major physiological changes stemming from the imbalanced karyotype. These include increased glucose metabolism (lowering of pH of the medium), a tendency to reduce contact inhibition (thereby leading to denser monolayers), and decreasing cell generation times from 18–24 to 10–14 hours.

In contrast to Dr. Moorhead's description of the limited life span of human euploid fibroblasts and the brief survival of Syrian hamster embryonic cultures, the majority of normal derivatives of the Chinese hamster are capable of continuous proliferation. This is "guaranteed" when aneuploidy develops.

It may be worthy to note, at this time, that control fibroblast cultures of the Armenian hamster (adult and embryonic) fail to survive beyond three passages unless transformed by some agent.

Spontaneous and induced neoplasia. During the past 14 years, over 100 tumors were processed for growth, *in vivo* and/or *in vitro*, with little or no success. Although limited survival *in vivo* may, in time, be improved by satisfying rigid histocompatibility factors, their nutritional requirements for culturing may remain obscured for some time.

In a recent sampling of 32 tumors (Table I), only 17 proliferated adequately to enable chromosome determinations. Ten were euploid, four were aneuploid, and the remaining three were pseudodiploid. Each of the seven variant tumors featured either a distinct and different marker chromosome, or an altered distribution of normal chromosome types, suggesting that each tumor mass stems from a single mutant cell, but one which may not have initially experienced the malignant transformation.

Aneuploid-pseudodiploid tumor derivatives grew more favorably than euploid forms during the first few passages. Many of the tumors which failed, either as primary or secondary cultures, were suspected of being essentially euploid. Although 50% of the tumors proliferated adequately for chromosome determinations, their growth rates fell drastically after the third passage, particularly among the euploids. Contrast this to the extended period of proliferation noted above for normal derivatives.

Altogether, 27 of the 32 tumors were implanted into random homologous hosts simultaneously at the time of culturing. Twenty-one tumors failed to yield even a single growth among 131 primary cheek pouch implantations. Six tumors propagated for 3–7 continuous transfers and had a combined tumor incidence of only 15% (26/171). The euploid tumor #89 can only be maintained *in vitro*. Even after experiencing aneuploidy-pseudodiploidy some 50 passages later, *in vivo* inoculations are still negative. In contrast, euploid tumor #127 could only be maintained *in vivo* for 20 transfers and had an overall tumor incidence of 30% (115/372). All twelve attempts to adapt this cell *in vitro* failed by the third passage.

Only one of the three initially euploid tumors (#89) remained stable for 50+ passages; the other two shifted to an aneuploid stem cell prior to the 20th passage. Five of the seven pseudodiploid-aneuploid variants also declined while maintained *in vitro*, with only one continuing for 100 passages, at which time it was discarded. By cyclic *in vitro* and *in vivo* propagation, re-cultured cells proliferated with renewed vigor, only to decline gradually, unless revitalized by reinoculating into animals. There was no apparent association between progression *in vivo* and aneuploidy, even though aneuploids fared better *in vitro*.

Three of the seven pseudodiploid-aneuploid tumor derivatives had deletion at the secondary constrictions of either both X's or chromosome 1.

A hamster-specific adeno-like virus was isolated from only one tumor (#89). After some 12 months of deliberate infection of cultures, the CPE pattern gradually

TABLE I

CHROMOSOME STATUS OF SPONTANEOUS AND INDUCED TUMORS OF THE CHINESE HAMSTER

Origin	Number of Tumors		Chromosome Studies on Successful Cultures		Chromosome Status					
					Euploid		Aneuploid		Pseudodiploid	
	Spont.	Ind.	Spont.	Ind.	Spont.	Ind.	Spont.	Ind.	Spont.	Ind.
Ovarian	6	6 x	1	3	1*	1*	0	1*	0	1
Uterine	3	3 x	3	3	2	3	1*	0	0	0
Pancreas	2	1 c	2	1	0	0	2*	0	1*	0
Adrenal	0	2 c	0	2	0	1	0	0	0	1
Subcutaneous	4	1 x	1	0	1*	0	0	0	0	0
Subcutaneous	0	4 m	0	1	0	1	0	0	0	0
Subtotals:	15	17	7	10	4	6	3	1	1	2
Totals	32		17		10		4		3	

* Established Cell lines.
c Cortisone (25 mg. subcutaneously).
x Whole-body X-irradiation (400r, 1–3 × 200r).
m 3-Methylcholanthrene (1 mg/0.1 Wesson Oil).

altered to one commonly noted for foamy agents (Yerganian, *et al.*, 1964; Yerganian and Nell, 1965). This exceptional tumor not only proliferated continuously *in vitro*, but is the only one from which viral agents have thus far been isolated.

Virus-transformed cells. Morphological transformation of embryonal derivatives with polyoma virus was evident 9–20 days following a moderate degree of CPE. With SV40 virus, focal proliferations were noted some 20 days after virus exposure. Seventy-nine out of 80 focal and clonal lines isolated from both cell systems prior to the third passage, or some 72 days after viral infection, were euploid. The one exception, clonal line B17, was pseudodiploid, and presumably this alteration was unrelated to virus activity since parent and other clonal lines were euploid. A range of morphologically different fibroblast- and epithelial-like clones were isolated from polyoma-treated embryonic cultures, while only transformed fibroblast-like elements were cloned from secondary cultures exposed to SV40 virus (Yerganian, 1963b). Euploidy persisted for some 30 passages among controls; the nine representative polyoma clones maintained continuously experienced aneuploidy during the 40th–85th passages; the SV40 fibroblast clonal subline was still euploid at the 125th passage. The spontaneous chromosome breakage frequency, covering some 20 passages, was $10 \times$ higher among control cultures (Yerganian, 1964). In contrast to normal and malignant derivatives of this species, transformed cells retained euploidy for much longer periods of rapid proliferation.

The only specific chromosomal disturbance noted during the first 72 days after exposure to polyoma virus was a distortion of the secondary constriction of both X chromosomes due to the persistence of nucleolar substance (Yerganian, 1963a; Yerganian, *et al.*, 1964).

Implantation of $3 \ 6 \times 10^6$ cells of three polyoma focal lines (3, 10, and 80) into cheek pouches (Table II) during the 3rd–14th passages, failed to produce a single tumor (0/12); collectively, six other euploid lines (9, A14, 22, 26, 41 and H) formed medium-sized growths in less than 30% of the inoculated cheek pouches (9/33); while the seven remaining lines (14, B17, 20, C24, A25, F25, and G25) formed large growths within 7–10 days in 40% of the cheek pouches (13/44). The latter three clones were derived from line 25 and were equally oncogenic.

When an aneuploid subline replaced the parent euploid population of line 9 by the 43rd passage, only one tumor formed among 14 cheek pouch inoculations. Similar trials with other cell lines, before and after the onset of aneuploidy, were also negative, suggesting that (a) the oncogenic potential is not influenced by minor increases in chromosome number or aneuploidy, *per se*, and (b) malignant expression is not a property of all morphologically transformed cell types.

After three successive *in vivo* transfers of a tumor stemming from line F14, cells were re-cultured and cloned. Twenty-two cheek pouch inoculations of a newly isolated clone (CP14) were made during the 21st–56th passages (right portion of Table III), with only eight growths (35%) appearing within 10 days. An XO subline emerged during this period and was cloned, to determine if the oncogenic potential is disturbed when the repressed X_2 chromosome is missing. The XO clone (#6 of CP14) formed tumors in only 20% of the cheek pouches, 15% less than the parent euploid cell.

TABLE II

ONCOGENICITY OF VIRUS-TRANSFORMED CELLS OF THE CHINESE HAMSTER

(3.6 × 10⁶ Cells per Cheek Pouch)

Trans. Gen.	Focal Line	No. of Implants	Growths	Chrom. Status	Trans. Gen.	Focal Line	No. of Implants	Growths	Chrom. Status
Polyoma									
3–14	3, 10, 80	12	0 *	Euploid	43–90	9	14	1	Eu-Aneu(XO)
3–14	9, A14, 22, 26, 41, H	33	9 *	Euploid					
3–14	14	2	2 [a]	Euploid	21–56	CP14 [a]	22	8	Eu-and XO
	20	7	2	Euploid					
	B17	5	2	Pseudo-	12–43	Clone 6 of CP14	14	3	XO
	C24	2	1	Euploid	15	CP17 [d]	4	0	Pseudo (X^m)
	A25	2	2	Euploid	43–101	B17	14	1	Pseudo (X^m)
	F25	9	2	Euploid					
	G25 [b]	6	2	Euploid					
SV40									
3–14	SV	7	1 [c]	Euploid	30–86	SV	12	7	Euploid
					43–45	TSV	5	4	Aneuploid
Control		—	—	Euploid	22–62	Control	9	0	Eu-Aneu.

* Medium-sized; others large.

[a] Three transfers in vivo, then re-cultured and cloned.

[b] Aneuploid sublines noted (very dense clones).

[c] Shifted to aneuploidy.

[d] Recultured after one in vivo transfer.

TABLE III

GROWTH POTENTIAL AND RELATIVE ONCOGENICITY OF PARENTAL EUPLOID AND ANEUPLOID SUBLINES OF DWARF HAMSTERS

Cell Type	Euploids			Pseudodiploids—Aneuploids			
	Growth	Oncogen	P.E. (%)	Growth	Oncogen	P.E. (%)	Cell Types †
Controls							
Chinese Hamster	+++	—	10	++++	—	50+	F
Armenian Hamster	—	—	0	—	—	—	F
Malignancies							
Spontaneous	±	±	15	++	±	15	F, I
X-ray and Cortisone	±	±	15	++	±	15	F, I
Polyoma	0	0	—	++++	±*	40+	F
A12 [a,b]	++++	+++	0	++++	++++	0	E
A18 [a]	++++	++	16	++++	++	40	F
In vitro trans.							
Polyoma	++++	+	12–40	++++	+	20–60	F, I, E
SV40	++++	+	37	++++	++	75	F
Parainfluenza 3	++	?	?	++++	0	?	E

† Fibroblast (F), Intermediate (I), and Epithelial-like (E).

[a] Armenian hamster; others, Chinese hamster.

[b] Whole-body x-irradiation—400R.

* Improved by Heterochromatin (Yerganian, et al., 1960).

The relation of the repressed X_2 to oncogenicity was again displayed while it was gradually being eliminated from the parent F14 line around the 58th passage. The population of F14 cells (based on metaphase counts) consisted of four sublines: (1) absence of repressed X_2; (2) presence of both arms of the repressed X_2 as functioning telocentrics, due to a transverse centromeric break; (3) either arm of the X_2 was present, and (4) parent euploids, which were now reduced to 4% of the population. A large tumor formed by the seventh day after inoculating a cheek pouch with approximately 4×10^6 cells. Upon reculturing the tumor, 95% of the metaphases were euploid, and all three forms of X-mutant cells were missing. The latter may have been represented by the small flecks of non-necrotic tissue encased by the fleshy euploid growth. Thus, the tumor stemmed from 16×10^4 euploid cells estimated to be in the inoculum. The restored parent euploid and mutant X_2 clonal lines remained stable until frozen-off during the 90th–95th passages.

The tumor incidence of the XO-transformed cells was 15% lower than concurrent trials with parent euploid cells. A similar relationship was noted following the loss of the Y chromosome among some aneuploid clonal sublines of a "spontaneous malignant transformation" (Yerganian, Leonard and Gagnon, 1961). Cultured cells exhibited extensive nuclear extrusions and budding (Longwell and Yerganian, 1965). These features still persist and may reflect the presence of an SV-like agent recently isolated from this cell type by Enders (personal communication). Earlier studies on transplantation of two polyoma-induced sarcomas (Yerganian, et al., 1960) suggested that when the amount of sex heterochromatin in clonal sublines exceeded the basic diploid ratio, both the incidence and tumor size were greater than those noted for clones with reduced amounts of heterochromatin. These observations are difficult to assess since we have yet to fully satisfy histocompatibility requirements.

Although aneuploidy per se failed to improve the oncogenic potential of polyoma-transformed cells, the SV40-transformed fibroblast line did improve with time (Table II). Its aneuploid (tumor) subline (TSV), isolated prior to the 15th passage, produced a higher incidence of tumors during the 43rd–45th passages.

Discussion

The relative order of euploidy among the various cell types can now be arranged in the following manner: virus-transformed > normal > neoplastic. Although aneuploidy per se does help to establish normal derivatives in vitro, the oncogenic potential of neoplastic and virus-transformed derivatives remains unaltered by minimal numbers of autosomal trisomies. Several unrelated instances have now been recorded where in vivo growth was noticeably reduced among clones bearing deletions of sex chromosomes. Conversely, in vivo growth of related clonal lines was improved when additional "repressed" chromatin was added to the karyotype in the form of modified X's or partial trisomies for either arm.

The majority of the initial aneuploid shifts in all cell types involve trisomy for chromosomes 5 and/or 8. The latter chromosome type participates in approximately

75% of the telomeric (end-to-end) associations, or sharing of nucleoli, with chromosomes 3, 4, and 6. *In vitro* proliferation is markedly improved by these trisomic states and may be due to an imbalanced production (or utilization) of nucleolar products, such as different families of nucleolar ribosomes, etc. The secondary constriction of the X chromosome is also associated with nucleolar activity. This site is prone to deletions, both spontaneously and following treatment with many different agents. The primary constriction of the X_2 is also readily fractured. Its heterochromatic arms may be deleted from the karyotype of cultured cells with no ill effects because of its repressed state. However, the significance of sex chromatin-associated nucleolar activity is witnessed only when dealing with *in vivo* growth of clonal sublines bearing different ratios of sex chromosomes/autosomes.

It is quite feasible to assume that during speciation, an active segment of chromatin bearing a prominent nucleolar organizer became translocated onto the functional segment of the neo-X chromosome. In so doing, the perinucleolar portions became repressed (position effect), while the nucleolar site continued to function and, perhaps, become accentuated because of its unique association as part of the sex mechanism.

In summary, it appears that *in vitro* proliferation is favored by additions of autosomes, preferably those having prominent nucleolar activity, while *in vivo* progression and oncogenic potential are influenced by corresponding increases or decreases in nucleolar sites linked to the X chromosome. Immunological studies are expected to help clarify these cytological trends. It has been established that the degree of chromosomal alterations noted in normal, malignant, and virus-transformed cells is strongly influenced by unknown species-specific factors. Retention of euploidy in these cell lines facilitates the separation of secondary alterations due to aneuploidy *per se* from those associated directly with the transforming principle of tumor viruses.

References

1. FORD, D. K., and YERGANIAN, G.: Observations on the chromosomes of Chinese hamster cells in tissue culture. J. Nat. Cancer Inst. 20, 393–425 (1958).
2. HSU, T. C., and ZENZIES, M. T.: Mammalian chromosomes *in vitro*. XVII. Idiograms of the Chinese hamster. J. Nat. Cancer Inst. 32, 857–869 (1964).
3. LONGWELL, A. C., and YERGANIAN, G.: Some observations on nuclear budding and nuclear extrusions in a Chinese hamster cell culture. J. Nat. Cancer Inst. 34, 53–69 (1965).
4. MOORHEAD, P. S.: Human tumor cell line with a quasi-diploid karyotype (RPMI 2650). Exptl. Cell Res. 39, 190–196 (1965).
5. PAPOYAN, S. A., YERGANIAN, G., ZILFIAN, V. N., and KHACHATUROVA, G. S.: The grey hamster as a new experimental animal in oncology. Proc. IXth Conf. Inst. Roentgen. and Oncol. Acad. Med. Sci. U.S.S.R. Yerevan, 71–73 (1964).
6. SACHS, L.: A theory on the mechanism of carcinogenesis by small deoxyribonucleic acid tumor viruses. Nature 207, 1272–1274 (1965).
7. YERGANIAN, G.: Transformation of Chinese hamster embryonal derivatives with polyoma and SV40 viruses. Genetics 48, 917 (1963a).
8. YERGANIAN, G.: Significance of variations in clonal morphology of polyoma and SV40 transformed euploid cells of the Chinese hamster. J. Cell Biol. 19, 93A (1963b).

9. YERGANIAN, G.: Transformation induced by polyoma and SV40 viruses in embryonal cell cultures of the Chinese hamster. Proc. Am. Assoc. Cancer Res. **5**, 70 (1964).

10. YERGANIAN, G., HO, T., and CHO, S. S.: Retention of euploidy and mutagenicity of heterochromatin in culture. In: *Cytogenetics of Cells in Culture*. R. C. J. Harris, ed., Academic Press, New York, pp. 79–96 (1964).

11. YERGANIAN, G., HO, T., and CHO, S. S.: Euploidy and chromosome alterations in normal, malignant and tumor-virus transformed cells of the Chinese and Armenian hamsters. Proc. Symp. on Mutational Process, Prague, 1965 (*in press*).

12. YERGANIAN, G., KATO, R., LEONARD, M. J., GAGNON, H., and GRODZINS, L. A.: Sex chromosomes in malignancy, transplantability of growths, and aberrant sex determination. In: *Cell Physiology of Neoplasia*. University of Texas Press, Austin, pp. 49–93 (1960).

13. YERGANIAN, G., and LEONARD, M. J.: Maintenance of normal *in situ* chromosomal features in long-term tissue cultures. Science **133**, 1600–1601 (1961).

14. YERGANIAN, G., LEONARD, M. J., and GAGNON, H. J.: Chromosomes of the Chinese hamster, *Cricetulus griseus*. II. Onset of malignant transformation *in vitro* and the appearance of the X_1-chromosome. Pathol. et Biol. **9**, 533–541 (1961).

15. YERGANIAN, G., and NELL, M.: Adenovirus CPE patterns in clonal cell lines of a spontaneous tumor of the Chinese hamster. Excerpta Medica **19**, 3 (1965).

16. YERGANIAN, G., and PAPOYAN, S.: Isomorphic sex chromosomes, autosomal heteromorphism, and telomeric associations in the grey hamster of Armenia, *Cricetulus migratorius*, Pall. Hereditas **52**, 307 (1965).

17. ZAKHAROV, A. F., KAKPAKOVA, E. S., EGOLINA, N. A., and POGOSIANZ, H. E.: Chromosomal variability in clonal populations of a Chinese hamster cell strain. J. Nat. Cancer Inst. **33**, 935 (1964).

Viruses and Human Neoplasm[1]

By

Joseph L. Melnick

Department of Virology and Epidemiology
Baylor University College of Medicine
Houston, Texas

The recent progress in understanding of viral transformation of cells *in vitro* and *in vivo* offers new avenues in the search for the viral etiology of human cancer. A few brief comments on some areas that were not included in the program of the conference, but which may be of importance for the human disease, seem to be in order.

Avian and murine leukemias induced by RNA viruses serve as one model for the study of *leukemia* in man. The advantages of using this model are manifold: (a) Infectious virus of distinct structure can be readily identified *in vivo* throughout the life span of the diseased animal, either circulating in the blood or in the malignant cells themselves. (b) Virus can also be detected throughout the life span of cultures that have undergone malignant transformation *in vitro*. (c) Virus can also be detected by its ability to interfere with the multiplication of related viruses in susceptible cells.

Recent studies on human leukemia have revealed findings similar to those observed in virus-induced leukemias of animals. Virus-like particles, similar to the avian and murine leukemia RNA viruses, have been found in the plasma or sera of a high proportion of leukemia patients, especially in children. Similar particles have also been observed in thin sections of bone marrow or lymph node cells of some patients with leukemia. The significance of these findings is still undetermined, for it has not been demonstrated as yet that these particles possess biological activity similar to the virus particles of the animal leukemias.

Recently a lymphoma peculiar for children in Central Africa has been described (Burkitt's lymphoma). Particles with the structure of herpesvirus have been detected in nuclei of lymphoma cells derived from the tumors and propagated as continuous cell lines. There is no indication as yet that the particles observed are infectious. Even if they prove to be one of the known herpesviruses or a new member of the family, the problem remains as to whether they are etiologically related to the malignancy. Similarly, infectious reovirus has been recovered from a number of Burkitt tumors. Findings of this nature exemplify the great need for caution in

---

[1] Supported in part by research grant CA-04600 from the National Cancer Institute, National Institutes of Health, United States Public Health Service.

attributing an etiological relationship to viruses, or other infectious agents, which may be mere passengers located in the tumor. Indeed numerous reports have claimed isolation of infectious cytopathic agents from human leukemia, either *in vivo* in mice, or *in vitro* in various tissue cultures. Upon further study it was found that the viruses isolated either were agents indigenous to the inoculated animal, or turned out to be mycoplasmas which accidentally contaminated the cell cultures inoculated with the test material.

Recent electron microscopic studies reveal the finding of yet another type of virus-like particle, namely particles indistinguishable from papovaviruses, in autopsy material from patients suffering from progressive multifocal leuko-encephalopathy (a disease superimposed on either chronic lymphatic leukemia or other diseases of the reticuloendothelial system). These observations also await further tests as to whether specific biologic activity accompanies the morphological findings of virus-like particles.

Another similarity between human leukemia and the virus-induced leukemias of animals, especially avian leukemia, comes from studies of cells derived from leukemia patients and studied in long term tissue culture. It had been observed that cells of various organs derived from chickens infected with avian myeloblastosis virus undergo a spontaneous conversion to myeloblasts when propagated *in vitro*. As with the experimental *in vitro* transformation of normal chicken target cells by this virus, the spontaneous proliferation of myeloblasts in cultures of infected organs is attributed to the transforming capability of the virus. Similarly spontaneous conversion to immature lymphoid cells has been found in fibroblastic cultures of bone marrow cells derived from children with acute leukemia. These lymphoid cells can be propagated in culture indefinitely and have been extensively studied. Lymphoid and myeloid cell lines have been derived from peripheral blood cells of patients with lymphocytic or with myelogeneous leukemias; in addition lymphoid cells derived from Burkitt's lymphomas have also resulted in permanent cell lines.

However, attempts to demonstrate a cytopathic, or interfering, or cell-transforming agent in any of these human leukemia cells have thus far failed. Thus, in contrast to the findings in virus-induced animal leukemias, in which infectious virus can be demonstrated with relative ease, work on human leukemia has been suggestive of viral causation but it has not been rewarding as to actual isolation of an active biological agent etiologically related to the disease.

There may be many reasons for this failure. First, there is the obvious drawback of the lack of a susceptible host for *in vivo* experiments. It is hoped that the current trials with newborn laboratory primates may yield results in the years to come. Secondly, the concentration of infectious virus in the naturally occurring disease of man may be very low. Thirdly, the tissue culture systems used to demonstrate the transforming capacity of materials derived from the patient or from cell lines grown in culture may not be sensitive enough. As shown with avian myeloblastosis virus, only specific target cells undergo transformation. Fourthly, if indeed human leukemia is caused by a virus, the responsible virus may be a defective one and special procedures may be needed to demonstrate its presence. Thus many avenues still remain open in the search for a viral agent in human leukemias and lymphomas, utilizing the models offered by the RNA tumor viruses of animals.

In view of the finding that, with few exceptions, *solid tumors* in animals induced by the adeno- and papovaviruses are free of infectious virus or of infectious nucleic acid, it is not surprising that innumerable attempts to isolate viruses from a variety of solid human tumors have met with complete failure.

The new knowledge of virus-induced T antigens in tumors (or in transformed cells) and of T-antibodies in the tumor-bearing animals has led to similar immuno-logical studies of cancer in man. Several attempts, unsuccessful as yet, have been made to test sera of cancer patients for the presence of complement-fixing or im-munofluorescence antibodies to possible T antigens in the patient's own tumor cells.

The finding that the DNA tumor viruses induce these T antigens, not only in the cells they transform but also in the early stages of cytocidal virus replication, offers a model for similar studies with known viruses of man not yet implicated in animal carcinogenesis. Cooperative studies using the sera of cancer-bearing patients are being planned with the support of the National Cancer Institute. Candidate viruses for such studies include not only the adenoviruses but also other viruses of man. For example, strong candidates include members of the herpesvirus group, such as simplex, zoster, and cytomegalovirus, all known to exist in a latent state in man for long periods of time.

A converse approach employs antisera of tumor-bearing animals that contain T-antibodies for the known tumorigenic viruses—those in papova and adeno groups. These antisera are being studied to determine if they will detect the presence of virus-induced T antigens in human tumors.

The techniques for searching for viruses in human cancer have become much more sophisticated than those currently used in the virus diagnostic laboratory. How-ever, investigators must still contend with the problem of identifying "passenger" viruses present in the cancer but not in a causal relationship. Of even greater con-cern is the converse, the problem of identifying a virus no longer present in the cancer that it may have caused. Some of the approaches that have proven fruitful in the study of malignant changes in cells and animals were discussed yesterday, and this morning the discussion will be focused on the significance of malignant transformation in relation to human neoplasm.

Transformation of Human Cells by Oncogenic Virus SV40

By

ANTHONY J. GIRARDI and FRED C. JENSEN

The Wistar Institute, Philadelphia, Pennsylvania

To date, the only viruses that have been shown capable of transforming human cells *in vitro* are Simian Virus 40 (13, 16, 20, 24) and a hybrid strain or adenovirus type 7-SV40 (strain LLE-46) (4). This latter virus is a human strain of adenovirus which has apparently acquired a portion of the SV40 genome during laboratory passage. By currently accepted standards, infection of human tissue with Rous sarcoma or wart viruses has not led to transformation.

Studies with SV40. A brief description of the uninfected cell strains employed in the studies to be discussed seems in order. The cultures were derived most frequently from human embryonic lung tissue, although adult skin and buccal mucosa reacted similarly following virus infection. The embryonic strains referred to in this report are the diploid cell strains, WI-26 and WI-38 (8). The cells were fusiform or fibroblast-like in morphology and grew in monolayers with some whorls during earlier passages where cell densities were greater. Confluent cell sheets developed in 3–4 days after subculturing, at which time cell mitoses all but disappeared. This general pattern of active growth (termed Phase II) was maintained until about the 50th passage when control substrains required 7 days or longer to achieve a confluent monolayer. Eventually, this downward trend in cell growth (designated Phase III), was followed by complete cessation of cell division.

SV40 infected cell cultures. The exact times at which each of the events, collectively termed transformation, occur, are influenced by several factors (discussed below). However, the same sequence of events is observed following infection of human cells by SV40.

First, there is a period of outward quiescence with no detectable change in growth rate, morphology, or cell karyotype, and contact inhibition of cell division is still operative. The first recognized change is the loss of this contact inhibition (13, 24). In human fibroblast cultures, cell division does not usually occur after the cultures become confluent. This effect is associated with monolayering, but properly should be distinguished from contact inhibition, a phenomenon of cell migration as defined by Abercrombie and Heaysman (1, 2). Stoker (23) has, in fact, indicated the distinction between inhibition of cell migration and of cell division. The loss of growth inhibition is followed by elevation of the cell-increase ratio, and sometime later, by an additional increase in growth due to a higher mitotic rate.

Examination of stained coverslip preparations for the earliest appearance of changes in cellular morphology revealed no consistent differences between control and infected material until several weeks after the loss of contact inhibition was detected. Thereafter, most of the cells assumed a more compact form, as has been observed in many previous studies.

Rapid cell proliferation was then sustained for several months following the initial transformation. Of course, during this time the uninfected control cells ceased dividing and the normal culture could no longer be subcultured. During the long period of rapid growth of the transformed culture, bizarre nuclear forms, as well as multinucleated cells, resulted from disturbances of the spindle and rearrangements of chromosomes. These became an increasingly prominent feature of the morphology of the transformed line, and cell growth was decreased markedly.

The decline in growth rate, which is characteristic of the overall process of transformation of human cells with SV40, is termed "crisis" (7) and is discussed at greater length below. Surviving cells, apparently dormant, emerged after several weeks or months and resumed the high growth rate characteristic of transformed cells. Recovered lines have been continuously cultivated and are presumably capable of indefinite growth *in vitro*.

The karyologic findings in human cells infected by SV40 were presented at this meeting by Dr. Paul Moorhead and have been published elsewhere (15).

Obviously, *sensitivities of detection* of various parameters for transformation may differ during the initial observation period following infection, so that the recognition of one event before another, such as the loss of contact inhibition prior to detectable chromosomal abnormalities, must be interpreted with reservation.

Virus-host relationships during transformation. Titration of culture fluids and extracts prepared from infected cells indicated that a moderate level of infectivity was maintained from the time of infection until well after the period of recovery from crisis. However, the number of cells shedding virus and, more important, the amount of infectious virus released per shedding cell depicted more accurately the nature of the host-virus relationship at different times after infection. For instance, in one study, at 14 days after infection the proportion of cells shedding infectious virus was quite low (0.13%), although these cells "released" an average of about 685 $TCID50$ per shedding cell. During the next 4 weeks, the percentage of cells releasing virus increased to approximately 100% while the virus yield per cell decreased until an average of only 1.6 $TCID50$ per cell was detected. A decrease in the percentage of cells shedding virus then occurred and a low level (1–5%) was maintained thereafter. The amount of virus shed per cell was also limited but persisted until crisis. Different sublines of the same cell cultures which recovered from crisis did not behave in the same manner with respect to *eventual* infectious virus production. For instance, one subline may continue to shed virus $1\frac{1}{2}$ years after its recovery as an established line, while another subline may cease producing virus. In another study, 1 of 10 recovered sublines shed virus, while 9 sublines have remained negative.

Viral antigen studies. The data pertaining to infectious SV40 correlated well with that collected from simultaneously performed immunofluorescent studies for *viral antigen*. In the example given above, the proportion of cells with positive nuclear staining was always less than 1% (day 14–day 189), and the viral antigen was difficult to detect even at the time when 100% of the cells were releasing infectious virus according to the infectious center test. What at first seemed a paradox was resolved when infectious virus yields per cell were examined and found to be an average of only 1.6 TCID50 per cell when all cells were virus-producers. This small amount of virus, plus the lack of fluorescent staining, indicated that at that time most cells were producing little or no viral coat protein able to react with sera containing SV40 neutralizing antibody. The most intense staining for viral antigen occurred at 2 weeks post-infection; this fact correlated well with the data for the high virus yield per cell which averaged 685 TCID50 per cell.

SV40 induced complement-fixation antigen. This antigen was detected by two methods; direct assay of cell extracts in the micro-CF test (10), and examination of cells grown on coverslips by the fluorescent antibody technique (17) employing specific serum obtained from hamsters bearing non-virus shedding SV40-induced transplant tumors. Using the fluorescent antibody method, the antigen was detected at all intervals studied. Initially, about 1% of the cells were reactive, and this value persisted until day 80 post-infection. Thereafter, the values increased until all cells were found to contain antigen. However, the level of antigen was too low to permit detection in the micro-CF test until day 90, but thereafter the values increased and remained detectable for the duration of the experiment.

Examination of the above data shows that an inverse relation between virus yield per cell and the number of cells positive for SV40 CF antigen developed during transformation. The relation of virus to its host cell apparently changed from one in which infectious virus was synthesized and shed at a relatively high level per cell, to one in which production of infectious particles was quite inefficient. This shift was followed by an increased proportion of cells showing the presence of SV40 CF antigen. All of the extensive and dramatic changes in growth, karyologic and morphologic parameters that define transformation appear during the time when this new host-virus balance is becoming apparent.

It has been shown (5, 18) that the SV40 CF antigen is present as an "early" antigen even in cells that eventually undergo lysis, and as the infection proceeds, the amount of this antigen increases temporarily and then may be reduced with concomitant synthesis of virus coat protein and virus maturation. In our study, in a sense, an "abortive" cycle of infection seems to occur in those cells which undergo transformation, since finally only the "early" antigen is being synthesized; further virus synthesis and maturation does not *usually* occur. Such a cryptic "infection" would have to be transmissible to daughter cells, but a true "integration" of viral DNA into the host genome may or may not be invoked.

Such a relationship with SV40 might be considered comparable to that which occurs in chick cells with the Rous sarcoma virus. Although the similarity between the two host-virus systems probably ends at this point, the important cell-virus relationship could very well involve an abortive or defective cycle of maturation. Rubin has suggested that the high carcinogenic potency and the defectiveness of RSV are associated in more than a casual way (19).

This postulated shift by the virus to some abortive or cryptic state in the cell eventually must be reconciled with the mutation-like properties that are conferred upon the transformed cell.

Factors which influence the rate of transformation. Previously, I mentioned that several factors may influence the times when the various events comprising transformation occur. As might be expected, there is a direct virus dose response effect which is expressed in several ways. Not only does the amount of virus employed per culture determine the transformation time, but factors that influence efficiency of virus adsorption are equally important. For instance, direct contact of the virus inoculum with the bare cell sheet during adsorption shortens the time to transformation.

More important is the age, or passage level, of the culture at the time of infection. This was described independently by Jensen (12) and Todaro (24). Cultures which are infected in Phase III, near the end of their finite lifetime *in vitro*, transform much more rapidly than do cultures infected during Phase II; e.g., 4 to 21 days vs. 35 to 70 days, respectively.

However, even this idea must be qualified since early passage Phase II cultures show variation in transformation time depending on whether they are infected as stationary or actively growing cultures. The latter, infected during the peak of mitosis, transform rapidly as do Phase III cultures; that is, loss of contact inhibition of division is detected in less than 3 weeks. This is an interesting finding since Phase III cultures might be considered as the extreme example of "stationary" cultures. The clue towards solving this paradox might lie in the similarity of the host-virus relationships which develop in the actively growing Phase II cultures and the "stationary" Phase III cultures. The abortive, non-infective type of relationship seems to occur more effectively in these two systems. I strongly suspect that this is an important step in the parasitism of host cells by oncogenic viruses, at least of the DNA type.

Another factor which influences the transformation time is the ability of normal cells to suppress the expression of transformed cells. This was noted by Stoker (23) and by Shein and Enders (20). In our studies, mixtures of SV40 transformed and normal cells of the same original strain were mixed and subcultivated at routine intervals. There was a definite lag before visual evidence of "transformation," and the time was not in keeping with that expected if transformed cells were growing at their usual rate. One is forced to consider this point when reviewing the data after infection of cells during Phase III: is rapid cell change due in part to the paucity of normal cells with inhibitory properties? On the other hand, why do SV40-infected, actively growing (not stationary) Phase II cells transform as well as Phase III cells even though most of the population appears to be normal? Todaro and Green (25) have described experiments which indicated that SV40 could not initiate transformation in a non-growing population of mouse 3T3 cells. The "fixation" of the transformed state in the genetic sense required one cell generation after infection and the expression of the transformed state required several more generations.

Crisis and recovery. The first results of studies of human diploid cell strains infected with SV40 gave an impression of transformation as a basically proliferative response. Although the infected cells survived propagation times for uninfected controls, it became evident that transformation involved more processes. Three to eight months after initial morphological alterations, the cells underwent increasingly abnormal mitosis, and subcultivation could not be continued readily. From this phase, called "crisis," through minimal cell survival, autonomous cell lines emerged with features associated with other continuously propagable culture lines. Shein (21) and Moyer (16) described a period following SV40-induced transformation similar to crisis, but only exceptionally observed the phenomenon of recovery as reported here.

The time relationship between Phase III in normal human diploid-strain cultures and crisis in the transformed cell lines. In the course of SV40 transformation studies of human tissues infected during Phase III, we observed that transformed cells survived for about 9 weeks *post Phase III of the control cells.* To determine whether crisis represented a "delayed Phase III," the following experiments were performed. Cells of strain WI-38 at culture passage 12 were infected with SV40. Uninfected controls were also carried and when these reached passage levels 20, 30, and 40, cultures were infected under identical conditions as had been used for P12 cultures, employing sister ampules of one virus pool. The 4 infected sublines transformed and the parent uninfected control line reached Phase III.

The lines infected at passages 12, 20 and 30, though transformed at different times, went into crisis at the same time; the line infected at P40 went into crisis one week later. The time interval between Phase III of the normal cells and crisis was 10 weeks. This is in agreement with past experiments where single lines resulting from infection at one point in a cell's life cycle were used in each study.

Attempts to circumvent crisis or to insure recovery of established lines. Attempts to circumvent crisis have failed. The use of SV40 immune serum as well as alteration of the pH of the medium or of the split ratio did not prevent crisis. However, certain important points have been established. Low split ratio (1:2) and low initial pH of the medium (pH 6.9) allowed cultures to survive for longer periods of time before the inevitable event. More important was the finding that sister cultures should be held without subcultivation, *at each level* when crisis is approaching, to attempt recovery.

From our results, it is probable that the infectious SV40 virus production and reinfection are not the cause of crisis but that the event is a controlled, biologically timed property of the cells represented by a "delayed Phase III."

Properties of cells recovered from crisis: On the property of infectious virus production (or non-production) of cells post-crisis. Sublines of SV40 transformed cells may behave differently after recovery. Some continue to shed infectious virus in relatively high titer, whereas others do not produce infectious virus at a detectable level. When several other properties of these 2 cell types were compared, no differences were noted. For instance, their times of recovery, growth rates, ability to produce SV40 induced complement-fixing antigen, lack of sensitivity to mixtures of SV40 immune serum plus complement, and general karyology were similar. Infectious center assay has indicated that a low percentage of cells in the positive cultures release virus *at any one time.* When subcultured for 3 months in the presence of SV40-immune calf serum and thereafter with ordinary calf serum, the cells continued to shed infectious virus. Therefore, it is possible that a few cells capable of shedding virus appear at each passage level, and that virus production is not perpetuated by reinfection. This *may* imply that all cells are capable eventually of infectious virus production. Current studies with clones established in the presence of SV40-immune serum may help to clarify this point.

Superinfection of sublines recovered from crisis. In one study, 10 sublines of one transformed line recovered from crisis. Only one of the ten continued to shed virus. The non-virus shedding cell lines were then treated with SV40. Seven of the nine sublines were superinfectible and two sublines were not. The superinfected lines were checked for presence of SV40 viral antigen by use of fluorescein labeled antisera *before and after infection*. The values for positive stained cells were 2–5% after infection and 0% before. The two sublines which were not superinfectible were negative. Six months after superinfection, infectious virus was still produced in low titers, $10^{1.5}$–$10^{3.5}$.

In addition to superinfection with SV40, the 10 sublines were examined for (a) production of interferon, (b) refractoriness to infection with poliovirus and vesicular stomatitis virus (VSV), and (c) capacity of the cells to transfer refractoriness to other cells (9). None of the sublines produced interferon as detected by tests with poliovirus type 2 and VSV. However, 2 sublines when challenged directly with these two agents were somewhat resistant to infection and one subline was markedly resistant to infection with either virus. These tests were repeated with similar results. The interference effect could not be transferred to human-embryo-kidney tissue cultures.

The above data indicates that, although the chronological events which constitute crisis and recovery seem to be similar for all sublines of different cell strains studied to date, certain properties may vary such as virus-shedding capacity, superinfectibility with SV40, and resistance to infection by other viruses.

Attempts to induce infectious SV40 or rescue SV40 genome from SV40 transformed non-virus producing cells. Attempts to induce non-virus shedding cells to yield SV40 by superinfection with other viruses have failed (adenovirus types 4, 7, 12, 18 and UV-irradiated vaccinia virus). The viruses which were recovered were then analyzed for their ability to induce production of SV40 CF antigen in cultures of green monkey kidney cells. None was produced in cultures inoculated with any of the virus strains. Also, the recovered viruses were inoculated into newborn hamsters. The induced tumors and transplants thereof will be analyzed for new induced complement-fixing antigens to determine whether SV40 information has been transmitted with the "carrier" virus.

With similar purpose, DNA extracts were prepared from transformed cell lines and from normal cells. These extracts were used to treat normal human cells in Phase II and Phase III. None of the cultures showed the usual signs of transformation after treatment with transformed cell DNA.

Development of antigens responsible for producing tumor resistance. SV40 virus-induced tumors and cell cultures transformed by the virus possess new antigens. The one most commonly studied has been the induced complement-fixing antigen first described by Black, Rowe and Huebner and their co-workers (3), although results of studies which measure parameters other than CF ability suggest that more than one new antigenic component is present as a result of virus-induced transformation. These changes are especially interesting because antigens of SV40-induced hamster tumors, or cell cultures prepared from such tumors, are highly effective in preventing the occurrence of tumors in hamsters inoculated with SV40 when newborn. Since results of studies with SV40 CF antigen suggest that antigen is located in the cell nucleus, it seems unlikely that the tumor rejection mechanism is directed against, or mediated by, this antigen. *Human cell cultures* transformed by SV40 similarly possess the CF antigen which is retained after the cultures have recovered from crisis and no longer shed infectious virus. It was of interest to determine whether SV40-transformed *human* cells could also prevent SV40 viral tumorigenesis even though the cells are of a species foreign to the experimental host. The evidence indicates that indeed this ability is present and probably results from an antigen other than the CF antigen (6).

In these studies, the inducing virus was inoculated into hamsters less than 1 day of age; virus-free human cell suspensions were injected intraperitoneally during the latent period preceding appearance of tumors. The prevention of virus-induced tumors was most effective when whole cells were used; frozen and thawed or formalin-treated preparations at equivalent concentrations did not reduce tumor incidence. Since the inactive frozen and thawed preparations contained SV40 complement-fixing antigen in high titer, another antigenic component was responsible for the protective effect.

Studies on SV40 tumor enhancement by sera from animals demonstrating tumor resistance. Sera from SV40 tumor-resistant animals were mixed with tumor cells and reinoculated into the *autologous* serum donors. Such inocula led to formation of tumors in the previously *"immune" host*. A control inoculum of tumor cells and normal hamster serum placed on the opposite flank of the same animal was rejected completely. These results were confirmed by similar studies in which the mixtures of tumor cells and "immune" or normal sera were injected into normal hamsters. In these animals, tumors could grow at both inoculation sites; however, the tumor cell-"immune serum" inoculum led to earlier tumor formation and enhanced growth.

It is possible, as suggested by Snell (22) and by Moller (14) that surface antigen(s) are coated by the enhancing antibody from sera from immune animals, and this leads to an afferent inhibition of the protective homograft reaction of the host. Our fluorescent antibody studies show that these enhancing sera in fact do coat the cells, so that this reaction may be due to the presence of surface antigens *related or unrelated* to the homograft-type antigen. Therefore, at this time it cannot be assumed that serum from animals which show resistance to tumor formation, and stain cells by the fluorescent antibody method, are staining for "tumor resistance antigen."

Homologous implantation of SV40 transformed human cells into terminal cancer patients. This study was conducted in cooperation with Drs. R. Crichlow and R. Ravdin at the Hospital of the University of Pennsylvania and results were published in part in a previous report (11).

Growth of a nodule was indicated by the appearance after 3–5 days of a discrete, firm mass at the site of implantation that increased in size over a period of several days. Fully grown nodules were hard and non-tender, rounded, elevated, and well demarcated, and occasionally the overlying skin was erythematous. Although at times a developing nodule was obscured by an early inflammatory reaction, continued examinations of the implantation site permitted its identification. Unless excised, nodules regressed in 10–28 days after implantation.

Detectable cell growth after implantation occurred with inocula as small as 1×10^6 cells, and in 50% of instances with inocula of 5×10^6 cells. Nodules were excised between 3 and 14 days and either fixed for histopathologic evaluation or grown *in vitro*. Fixed sections of biopsied nodules showed proliferating neoplastic cells with an accompanying inflammatory reaction predominantly lymphocytic in later biopsies. Cells recovered and reestablished in tissue cultures were examined for various properties and found similar to those implanted by morphology, growth rate, and more important, the presence of SV40 induced CF antigen. The latter could be detected in 100% of the cells by staining the primary cell cultures growing out from nodule biopsies or, when sufficient numbers of cells were available several passages later, by direct measurement in the CF test. Also, chromosomal examination of cells recovered from biopsy revealed that the range of karyotypes was consistent with those of the cells implanted. This type of analysis, especially where a "marker" chromosome was present, would indicate that the cells were recovered from the transformed cells which had been implanted.

Homotransplantability was acquired in the stage of transformation *prior* to crisis, although the inoculum size had to be increased to cause production of a nodule and the incidence of positive takes was less than following implantation of post-crisis cells.

In vitro transformation of human cells by adeno 7-SV40 hybrid virus. Black and Todaro (4) have reported the transformation of human fibroblasts by an adeno 7-SV40 hybrid, strain LLE-46. It was stated that the available evidence indicates that a portion of the SV40 genome is enclosed in the adeno-7 protein coat and may be responsible for the transforming activity. The transformations of hamster and human cells by LLE-46 are in most ways similar to those described with SV40; the morphologic changes, the growth characteristics, including the period of crisis, and the chromosomal aberrations were likewise similar. Also, the SV40 neoantigen was present in most cells. One difference was the shorter latent period for transformation with the hybrid virus at comparable multiplicities of infection. The transformation could be inhibited by adeno-7 antiserum but not by SV40 antiserum.

Studies with "ordinary" viruses for their ability to transform cells in vitro have been limited, due to the cytopathic effects (CPE) of the agents in the systems used for testing. However, one may take advantage of primate cells, which are relatively insensitive to CPE for selected agents. Tests for transforming ability may be performed by employing dilutions of virus which contain abundant infectious particles but which do not cause CPE. For example, studies have been performed with adenovirus type 7 (not a hybrid with SV40 genome) in African green monkey kidney cells (Jensen and Girardi, in preparation). Although the virus titer in a sensitive titration system was 10^{-7}, the agent did not cause CPE in the GMK cultures at dilutions above 10^{-2}. At these non-CPE dilutions the agent could induce changes in the host GMK cultures which could be considered transformation. The cells had altered morphologic features, increased growth rate, displayed loss of contact inhibition and demonstrated the presence of adeno-7 induced complement-fixing antigen.

If this method is applicable to other agents in human and monkey tissue, it may represent an avenue of approach to the study of the transforming ability of ordinary viruses indigeneous to the human population.

References

1. Abercrombie, M., and Heaysman, J. E. M.: Observations on the social behavior of cells in tissue culture. I. Speed of movement of chick heart fibroblasts in relation to their mutual contacts. Exper. Cell Res. **5**, 111–131, 1953.
2. Abercrombie, M., and Heaysman, J. E. M.: Observations on the social behavior of cells in tissue culture. II. "Monolayering" of Fibroblasts. Exper. Cell Res. **6**, 293–306, 1954.
3. Black, P. H., Rowe, W. P., Turner, H. C., and Huebner, R. J.: A specific complement fixing antigen present in SV40 transformed cells. Proc. Nat. Acad. Sci. **50**, 1148–1156, 1963.
4. Black, P. H., and Todaro, G.: *In vitro* transformation of hamster and human cells with adeno 7-SV40 hybrid virus. Proc. Nat. Acad. Sci. **54**, 374–381, 1965.
5. Gilden, R. V., Carp, R. I., Taguchi, F., and Defendi, V.: The nature and localization of the SV40-induced complement fixation antigen. Proc. Nat. Acad. Sci. **53**, 684–692, 1965.
6. Girardi, A. J.: Prevention of SV40 virus oncogenesis in hamsters. I. Tumor resistance induced by human cells transformed by SV40. Proc. Nat. Acad. Sci. **54**, 445–451, 1965.
7. Girardi, A. J., Jensen, F. C., and Koprowski, H.: SV40-induced transformation of human diploid cells: Crisis and recovery. J. Cell. Comp. Physiol. **65**, 69–83, 1965.
8. Hayflick, L., and Moorhead, P. S.: The serial cultivation of human diploid cell strains. Exp. Cell Res. **25**, 585–621, 1961.
9. Henle, G., Henle, W., and Girardi, A. J.: *Unpublished results.*
10. Huebner, R. J., Rowe, W. P., Turner, H. C., and Lane, W. T.: Specific adenovirus complement fixing antigens in virus-free hamster and rat tumors. Proc. Nat. Acad. Sci. **50**, 379–389, 1963.
11. Jensen, F. C., Koprowski, H., Pagano, J. S., and Ravdin, R. G.: Autologous and homologous implantation of human cells transformed *in vitro* by simian virus 40. J. Nat. Cancer Inst. **32**, 917–937, 1964.
12. Jensen, F., Koprowski, H., and Ponten, J. A.: Rapid transformation of human fibroblast cultures by simian virus 40. Proc. Nat. Acad. Sci. **50**, 343–348, 1963.
13. Koprowski, H., Ponten, J. A., Jensen, F., Ravdin, R. G., Moorhead, P., and Saksela, E.: Transformation of cultures of human tissue infected with simian virus, SV40. J. Cell. Comp. Physiol. **59**, 281–292, 1962.
14. Moller, G.: Studies on the mechanism of immunological enhancement of tumor homografts. I. Specificity of immunological enhancement. J. Nat. Cancer Inst. **30**, 1153–1175, 1963.
15. Moorhead, P. S., and Saksela, E.: The sequence of chromosome aberrations during SV40 transformation of a human diploid cell strain. Hereditas **52**, 271–284, 1965.

16. MOYER, A. W., WALLACE, R., and COX, H. R.: Limited growth period of human lung cell lines transformed by simian virus 40. J. Nat. Cancer Inst. **33**, 227–236, 1964.

17. POPE, J. H., and ROWE, W. P.: Detection of specific antigen in SV40-transformed cells by immunofluorescence. J. Exp. Med. **120**, 121–128, 1964.

18. RAPP, F., KITAHARA, T., BUTEL, J. S., and MELNICK, J. L.: Synthesis of SV40 tumor antigen during replication of simian papovavirus (SV40). Proc. Nat. Acad. Sci. **52**, 1138–1142, 1964.

19. RUBIN, H.: Virus defectiveness and cell transformation in the Rous Sarcoma. J. Cell. Comp. Physiol. **64** (suppl. 1), 173–179, 1964.

20. SHEIN, H., and ENDERS, J.: Transformation induced by simian virus 40 in human renal cell cultures. I. Morphology and growth characteristics. Proc. Nat. Acad. Sci. **48**, 1164–1172, 1962.

21. SHEIN, H. M., ENDERS, J. F., PALMER, L., and GROGAN, E.: Further studies on SV40-induced transformation in human renal cultures. I. Eventual failure of subcultivation despite continuing high rate of cell division. Proc. Soc. Exper. Biol. Med. **115**, 618–621, 1964.

22. SNELL, G. D.: The suppression of the enhancing effect in mice by the addition of donor lymph nodes to the tumor inoculum. Transpl. Bull. **3**, 83–84, 1956.

23. STOKER, M.: Regulation of growth and orientation in hamster cells transformed by polyoma virus. Virology **24**, 165–174, 1964.

24. TODARO, G. J., WOLMAN, S. R., and GREEN, H.: Rapid transformation of human fibroblasts with low growth potential into established cell lines by SV40. J. Cell. Comp. Physiol. **62**, 257–265, 1963.

25. TODARO, G. J., and GREEN, H.: Cell growth and the initiation of transformation by SV40. Proc. Nat. Acad. Sci., USA. **55**, 302–308, 1966.

Discussion

BY ALAN S. RABSON

Pathologic Anatomy Branch,
National Cancer Institute,
Bethesda, Maryland

I would like to congratulate Dr. Girardi and his associates on the excellent work they have reported on transformation of human cells *in vitro* by SV40. Our own work with human cells has been much more limited and we have been concerned chiefly with the problem of whether the transformed cells have acquired neoplastic properties.

In our first experiments with SV40, we infected explant cultures of normal adult human thyroid tissue with the virus. The control cultures grew poorly while the infected cultures went through a "silent period" followed by rapid growth with the development of the morphologic and growth characteristics which Dr. Girardi has described in his transformed cells before "crisis." We made a number of attempts to produce tumors with these cells by injecting them into the cheek pouches and brains of radiated and cortisone-treated hamsters but all of these were unsuccessful. Limited growth of the cells was observed in the brains of cortisone-treated African green monkeys after intracerebral inoculation of large numbers of cells (1). Although the donor of the original tissue was alive, we did not believe autologous implantation experiments were justified and we decided to study the problem in animals.

Experiments similar to those with the human thyroid tissue were done with newborn hamster kidney explants, and again we observed transformation in the SV40 infected cultures (2). When cell suspensions from these cultures were injected into newborn hamsters, they rapidly produced tumors at the sites of injection. Histologically, in contrast to sarcomas producd by hamster kidney cells transformed *in vitro* by polyoma virus, the tumors were predominantly carcinomas with areas of adenocarcinoma and with tubule formation.

Thus, although SV40 transformed hamster cells are able to produce progressively growing neoplasms when inoculated into hamsters, and SV40 transformed human cells have morphological and growth characteristics similar to those of SV40 transformed hamster cells, the question of whether the transformed human cells are truly neoplastic remains. We, therefore, thought that it would be of interest to carry out transformation and cell implantation studies in sub-human primates such as monkeys to determine whether the findings in hamsters could be reproduced. To avoid histocompatibility problems when the transformed cells were inoculated *in vivo*, the tissues for transformation were removed surgically and the monkeys kept alive for subsequent autologous implantation of cells.

Explant cultures of muscle, skin and testis from three rhesus monkeys were infected with SV40. After several months, infected cultures grew more rapidly than non-infected control cultures and there were changes in cell morphology and growth characteristics similar to those seen in SV40 transformed human cell cultures. The transformed cultures were all virus-carrier cultures and SV40 virus was recovered from their supernatants during the first 27 months of the experiment. SV40 T-antigen was found in the nuclei of all the transformed cells by immunofluorescence (experiments carried out in collaboration with Dr. Richard A. Malmgren). Autologous implantation of as many as 10^8 cells failed to produce tumors in the animals (3).

The failure of autologous implantation of large numbers of transformed cells to produce tumors may be related to the presence of viral antigens in the cells. As Drs. Habel and Defendi have described, virus-associated transplantation antigens are present in the cells of hamster tumors produced by polyoma virus and SV40. Thus, although our SV40 transformed monkey cells acquired morphologic features and *in vitro* growth characteristics generally associated with neoplasia, they probably also acquired new antigens associated with SV40 infection which enabled even the autologous host to recognize them as "foreign" and to reject them.

References

1. Rabson, A. S., Malmgren, R. A., O'Conor, G. T., and Kirschstein, R. L.: Simian Vacuolating Virus (SV40) Infection in Cell Cultures Derived from Adult Human Thyroid Tissue. J. Nat. Cancer Inst. 29, 1123–1145 (1962).
2. Rabson, A. S., and Kirschstein, R. L.: Induction of Malignancy *in vitro* in Newborn Hamster Kidney Tissue Infected with Simian Vacuolating Virus (SV40). Proc. Soc. Exp. Biol. Med. 111, 323–328 (1962).
3. Rabson, A. S., Kirschstein, R. L., and Legallais, F. Y.: Autologous Implantation of Rhesus Monkey Cells "Transformed" *in vitro* by Simian Virus 40. J. Nat. Cancer Inst. 35, 981–991 (1965).

Immunologic Attack on Neoplasia[1]

By

M. R. Hilleman

Division of Virus and Cell Biology Research,
Merck Institute for Therapeutic Research,
West Point, Pennsylvania

In the simplest view, protection against cancer involves two different possibilities: first, protecting the host against those invaders from without which initiate the neoplastic process and second, protecting the host against the invaders from within which are the cancer cells themselves. The immunologic approach affords the means for attack both upon the invaders from without and the invaders from within. For convenience the immunologic approaches may be divided into the causative agent-specific, the tumor antigen-specific, and the nonspecific.

Causative Agent-Specific Control

On theoretical grounds, the most positive approach to control of cancer is to prevent contact between the host and those agents which cause cancer. In the case of the chemical and physical carcinogens, the approach is limited to avoidance of contact or exposure. In the case of carcinogenesis by viruses, there is wider possibility in that the immunologic control mechanisms of the host may be utilized for preventing or limiting the infection. The usefulness of the approach to control of cancer by preventing viral infection must necessarily depend upon the extent to which viruses are the cause of cancer. In man, carcinogenesis by virus remains a theory (1–19). The virus-like structures observed in human neoplastic tissues and in plasma have not been shown to be causally related to neoplasia. Some of the purported viruses isolated from cancer have been found to be mycoplasma and the mycoplasma, in turn, remain to be shown to have a causal relationship to cancer. True viruses or virus-like agents isolated from neoplastic tissues or from cancer cases have failed to date to be established as related etiologically to the malignancy.

The absence of data providing a definitive relationship between a virus and cancer in man precludes any serious attempt at cancer prophylaxis by viral vaccines at the present time. Hence, work on antiviral vaccines is justified only by presumption and is confined to exploration of basic principles in animal models.

[1] Research on cancer in our laboratories is supported in part by Contract PH43-64-55 from the National Cancer Institute, National Institutes of Health, United States Public Health Service.

Hope for effective viral vaccines appears limited to prevention of infection with cancer virus that is spread horizontally following birth. There is little basis for antiviral prophylaxis in vertically transmitted infections since the virus will already have had contact with the cells and immunologic tolerance will have had the opportunity to develop against both virus and transformed cells. Vaccines would be expected to be most effective against infection by oncogenic virus contracted later than the first several months following birth, at a time when the immunologic mechanisms of the body have been developed fully. Experience with animal tumors has shown, however, that infection with the tumorigenic agent prior to development of immunologic competence may be the common essential factor for cancer in the majority of virus-host systems. For such situations, the optimal approach might be to immunize the mother so as to provide a sufficient level of antibody in the newborn to protect against natural infection until the time of development of immunologic competence. Failing this, specific hyperimmune globulin administered early *post-natum* might achieve the same effect.

Vaccination against the virus itself would be useful principally as a prophylactic measure. In the case of virus-dependent neoplasia, however, some therapeutic benefit might be achieved by application of vaccine or specific immune serum in the hope of preventing infection of new cells or of damaging the neoplastic cells which contain virus. Whether for prophylactic or therapeutic application, live or killed vaccines might be considered. Live virus vaccines must necessarily be deprived of their oncogenic quality. Killed vaccines generally require several injections and immunity may not be lasting. Use of adjuvants with killed vaccine aid in boosting the height and duration of immune responses, protect against the suppressive effect of maternal or other circulating antibody, and stimulate maturation of the immunological mechanisms. Experiments aimed at evaluating the utility of the vaccine approach to control of cancer in animals have been limited to date. There are, however, sufficient examples to establish the utility of the approach of vaccination against infection with carcinogenic viruses in several kinds of circumstances. These are discussed briefly below.

Avian lymphomatosis. Avian lymphomatosis is illustrative of virus-induced cancer in which there is dependence on infection at an early age and in which immunologic tolerance is prominent. The infection may be transmitted vertically from infected dams via the egg or horizontally by contact of newly hatched susceptible chicks with secretions and excretions of infected chicks. Vertical transmission of the infection via the egg renders the chick immunologically tolerant to the virus with persistent viremia, virus infection of eggs laid by the bird in adulthood, and tumor development. Horizontal transmission generally requires contact with virus at an early age if persistent infection and tumor are to develop. Infection at a later age usually results in an adequate immune response, failure of virus excretion into the egg, and no tumor.

Burmester and associates (20) were able to protect against visceral lymphomatosis in virus-exposed chicks by previous vaccination of the corresponding dams with live fully virulent virus or with killed virus vaccines in aqueous form or in Freund's incomplete adjuvant. Fink and Rauscher (21) protected two-month-old

chicks against challenge with homologous Rous virus by administering formalin-
or heat-killed Rous sarcoma virus in Freund's adjuvant.

Chicks derived from dams with high level antibody against visceral lymphomatosis
virus are generally free of the agent and, in addition, are protected for a period of
time after hatching against infection by the passive antibody which is present.
Hughes and co-workers (22) utilized this fact in producing leukosis-free chicken
flocks by selecting eggs for hatching from dams which were high in antibody and
by rearing the progeny in isolation to prevent infection from without. There is no
direct evidence to suggest that avian lymphomatosis virus eaten in infected eggs or
injected with live virus vaccines causes cancer in man. However, there is concern
for this on theoretical grounds, since certain strains of the related Rous sarcoma
virus cause tumor in the newborn of a number of mammalian species and may cause
transformation of human cells in culture (23). Eventual finding of an association
of avian lymphomatosis virus with human neoplasia might necessitate development
of large-scale commercial production of leukosis-free eggs and chickens. The available
data provide a scientific rationale for achieving this objective, viz., by using eggs
from dams with high level antibody for hatching purpose and by immunizing the
offspring with live or killed virus vaccine at perhaps 3 to 6 months of age.
Immunization should prevent exogenous infection of the dams during the egg
laying period and should provide passive protection for the next generation until
such time that the progeny are immunized. Mass rearing of virus-free nonimmunized
chicks in strict isolation housing seems economically impractical as of now.

Mouse leukemia. The murine leukemias appear to be of the virus-dependent
type caused by RNA viruses resembling Myxoviruses. Type A leukemia viruses
require infection at early age to induce tumors but the Friend-Rauscher viruses are
carcinogenic even for adult animals. The Friend-Rauscher agents, therefore, lend
themselves to studies of prevention of neoplasia by virus acquired later than
development of immunologic competence. With these agents, it has been shown
(24–27) that killed virus vaccines given in aqueous or adjuvant preparation
stimulated a neutralizing antibody response and protected against infection and
induction of cancer upon later challenge with the virus. Prophylactic administration
of immune serum afforded similar protection and therapeutic administration of
antibody appeared to reduce viremia in infected animals.

Adenoviruses and papovaviruses. Adenoviruses are a frequent cause for epidemic
acute respiratory illnesses in man and such illnesses can be readily controlled by use
of killed or live virus vaccines. Types 3, 7, 12, 14, 16, 18, 21 and 31 and possibly
other adenoviruses of human origin have been found oncogenic for hamsters but
their carcinogenic potential for man is unknown. Because of this oncogenicity,
adenovirus vaccines are presently prohibited in the United States even though they
might perchance be good anticancer vaccines.

Eddy and co-workers (28) have shown that administration of killed polyoma
virus vaccine to female hamsters affords near complete passive protection against
the same oncogenic virus in the newborn progeny. It was also shown (29, 30)
that injection of a large amount of SV40 virus into adult hamsters which had
received the same virus when newborn prevents tumor formation. This seeming

paradox is most readily explained on the basis of immunity directed against virus-induced cell antigens and not against the virus itself. The large amount of live vaccine virus administered after immunologic competence develops in the animals appears to cause production of large numbers of malignant transformed cells of altered antigenic composition which are effective in stimulating immunity not only against themselves but also against the similar neoplastic cells transformed by the virus given initially.

Infectious myxomatosis. Infectious myxomatosis presents the example of direct use of an attenuated live virus vaccine. Saito and associates (31) were able to attenuate myxoma virus by passage in cell cultures of rabbit kidney. The vaccine was highly effective in inducing homologous neutralizing antibody and in providing durable immunity against challenge with fully virulent myxoma virus.

Antiviral drugs. Though nonimmunologic, the antiviral chemotherapeutic approach to control of cancer deserves mention. In animal tumor model systems, iododeoxyuridine has been shown effective in suppressing induction of adenovirus 12 tumors in hamsters when the drug was given shortly following virus (32). Rous sarcoma virus, though of RNA type, has at least one phase of replication associated with DNA and can be inhibited by drugs such as actinomycin D, mitomycin C, iododeoxyuridine and bromodeoxyuridine. Halogenated analogues of nucleosides suppress tumorigenesis by polyoma virus in hamsters (33) but it is not clear whether the drug action is antiviral or antitumor cell.

In the event of development of effective antiviral drugs, suppression of RNA virus might be useful not only in preventing infection by RNA tumor viruses, but also by prevention of further spread of infection in the host, and in the direct antiviral action in virus-dependent tumor cells. Likewise, inhibitors of DNA, such as the halogenated pyrimidines, might be useful in suppressing the activity of hypothetical DNA incorporated in the cell genome of virus-initiated tumors. Cancer chemotherapy, in general, should consider the need to destroy infectious virus in the cancer-bearing host as well as to destroy the tumor cells themselves.

Tumor Antigen-Specific Control

The abundant findings in experimental investigations of cancer have gone far toward establishing the role of tumor antigen-specific mechanisms in cancer but have not yielded any immediate technology which can be applied to prophylaxis or therapy of cancer in man. The approaches to tumor antigen-specific control of cancer might theoretically include both active immunization with tumor antigen and passive immunization using immune serum or active lymphoid cells from previously immunized hosts.

Whether active or passive immunization be attempted, the availability of appropriate antigen is the prime requirement for both approaches. If the specific antigens concerned with protective immunity in human neoplastic tissues are as diverse as in the chemically induced neoplasms of animals, then practicable prophylactic immunization would appear to be outside the question. If, however, a sizeable proportion of human tumor were associated with a single antigen, as might occur with widespread carcinogenesis by a single virus, preventive immunization would prove more feasible. Therapeutic immunization, irrespective of the degree of antigenic diversity, shows the greatest promise for the tumor antigen-specific approach since it is possible for the tumor-bearing host to supply the necessary antigen for his own vaccine.

Discussion of the tumor antigen-specific control of neoplastic disease will be limited to two examples: First, the application of autogenous vaccine in the autochthonous human host and second, the prophylaxis of virus-induced tumor in an animal model which we developed in our laboratories.

Immunization with autogenous vaccine. Surgical removal of a primary tumor provides a source of tumor antigen which can be processed into vaccine directly if in sufficient amount. If insufficient, it may be propagated in cell culture to increase the supply. Vaccination experiments with autogenous tumor vaccines have been carried out in man for more than a half century with variable and mostly negative results. The most extensive investigations were recorded by the Grahams in 1962 (34). These workers vaccinated 232 advanced cancer patients with tumor deoxynucleoprotein or live cancer cells of autogenous origin in Freund's complete adjuvant. One hundred thirty-nine patients served as unvaccinated controls. The vaccines failed to provide sufficient benefit to recommend their use in treatment.

Placental choriocarcinoma offers greater theoretical chance for immunologic control than other tumors do because it arises from the trophoblast and contains antigens corresponding to the father—thus partially allogeneic. Various attempts (35–37) at therapeutic immunization have been carried out using the paternal skin, sperm, leukocytes and the like as immunizing antigen. In a single case, remarkable improvement was reported. Tumor development, in spite of such immunologic difference between tumor and host, might be related to the rapid growth of choriocarcinoma but strongly suggests also that there might be immunologic tolerance against choriocarcinoma antigens.

Though autogenous tumor vaccines have shown little if any benefit to date, they still present the most likely approach to therapeutic treatment of neoplasia by tumor-specific immune mechanisms. The best hope for a therapeutic effect ought to be obtained by vaccination shortly following surgery so as to bring about immunologic rejection of distant metastases when they are still small enough in size to permit a positive effect. Such measures used alone are probably insufficient to achieve beneficial results and exploration of simultaneously applied measures intended to increase antigenicity of the tumor or to enhance the capacity of the host to respond immunologically ought to be encouraged.

Immunization after virus but prior to tumor appearance in an animal model. For purpose of studies of prophylaxis of virus-induced cancer, our laboratories (38, 39, 40) developed an animal model which might approximate the conditions for hypothetical neoplasia in man caused by viruses. In this model, hamsters were injected with SV40 or polyoma virus when newborn. Using a small virus dose, the latent period for first tumor appearance following virus was 98 and 34 days respectively and this was adequate to permit immunization with homologous hamster tumor antigen in the interim period.

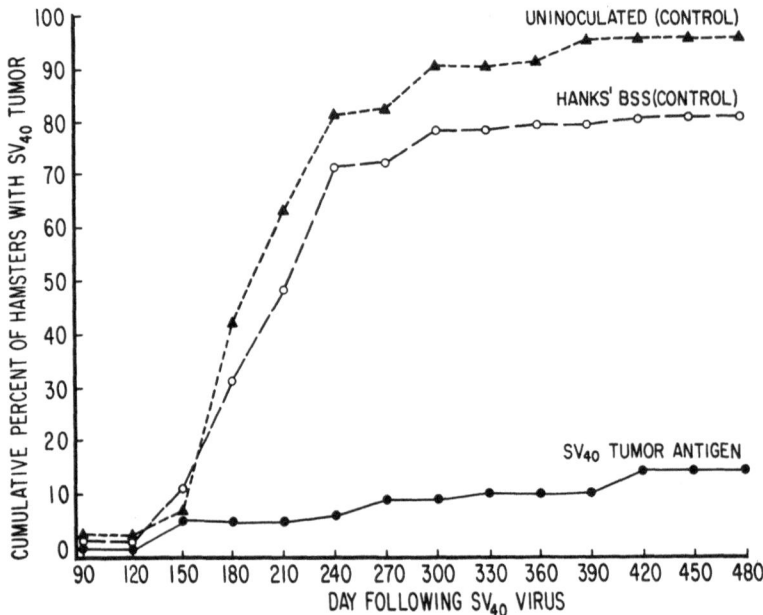

Fig. 1. Protective efficacy in hamsters of x-irradiated SV40 tumor cell antigen administered on day 34 following SV40 virus.

First attempts at immunization with dried and ether-extracted tumor antigen gave negative results (38). In a second series of experiments, hamsters which had received SV40 virus as newborn were vaccinated with x-irradiated homologous hamster tumor brei or x-irradiated tumor cells grown in culture (39). Typical findings with SV40 antigen are shown in Figure 1. The x-irradiated antigens were highly effective in preventing appearance of tumor when the antigen was given in a single dose as early as 34 days and as late as 76 days after SV40 virus was administered. Protective efficacy ranged from 3- to 7-fold. Girardi (41) has recently confirmed these data and has shown protective efficacy in the same model system using SV40 transformed human cells for vaccine purpose. The continued studies in our laboratories [1] have been directed toward refining the tumor preparation to eliminate normal antigens which might evoke autoimmune disorders and render the vaccines safe. All protective efficacy was lost, however, when the cells were disrupted in any way and the immunizing capacity was not preserved by prior treatment with formalin.

The efforts toward control of cancer by immunization with tumor cell antigens are thwarted not only by past history of failure but also by the possible danger it holds, especially if applied to persons with significant life expectancy. Any attempt to immunize an individual with preparations containing normal isoantigens or other antigens of autochthonous origin carries the possible hazard of inducing immune reactions against normal as well as tumor antigens with resultant autoimmune

---

[1] Drs. V. Larson, J. Coggin and M. R. Hilleman.

disorder. This danger is relatively small when aqueous vaccines are employed but may become very great when the antigens are administered in Freund's mineral oil adjuvant, especially with added Mycobacteria. Yet, the best hope for obtaining an adequate immune response to a weak antigen is by use of such adjuvant. An extremely important area for future research in cancer lies in the attempt to develop methods for isolation and concentration of tumor-specific antigens free of normal tissue antigens. The same danger from autoimmune disease may follow passive immunization with immune serum or sensitized lymphoid cells prepared in allogeneic or xenogeneic hosts. Transfer of competent lymphoid cells carries the added hazard of inducing graft-versus-host reaction. Cognizance must also be taken that immunologic enhancement rather than immunologic suppression may result from immunization. Added to these hazards are the problems of attempting practicable prophylactic immunization where multiple immunologic types of tumor are concerned, the difficulty of inducing effective immunity in an immunologically impaired or deficient individual, the problem of effecting immunologic response when the antigenic difference in tumor is only slight, the need for immunologic reactivation in cases of immunologic tolerance, and the chance for inadvertent transfer of cancer virus to normal persons given cancer vaccines prophylactically.

Nonspecific Control

In its broadest aspect, immunity includes all those processes both specific and nonspecific which are available to the individual host to resist or to overcome invaders from without or from within. Nonspecific defense mechanisms are of diverse variety and include physical barriers, inflammation, antimicrobial enzymes, fever, nonspecific humoral factors, interferon and the like. Present discussion will be limited to consideration of those factors which have been investigated extensively in relation to cancer or which merit possible consideration for the future.

The demonstrated role of the reticuloendothelial system in protecting against infectious disease agents led to the concept that cancer growth might reflect deficiency of reticuloendothelial function. Attempted stimulation of the reticuloendothelial system using microbes or microbial products such as BCG, zymosan, lipopolysaccharides, polysaccharides of diverse origin (Coley's toxins, Shear's polysaccharides), Freund's adjuvant, or antireticular cytotoxic serum has not been encouraging. Woodruff (42) has attempted passive transfer of lymphoid cells of xenogeneic or allogeneic origin to tumor bearing animals which had been treated with x-irradiation or cytotoxic drugs in nonlethal amount to destroy their own immune systems (adoptive immunity). The immunologically competent cells of the donor hosts appeared to afford some beneficial activity against the neoplasia in the recipient, but graft-versus-host reaction was a serious problem. No significantly beneficial effects were obtained when the method was carried to patients with advanced cancer. Mathe et al. (43) claimed some benefit of such adoptive immunotherapy in patients with leukemia who survived the graft-versus-host reaction.

Specific immunologic tolerance results from exposure to antigen prior to development of immunologic competence and also during immunologic competence if adequate dosages of antigen are used. Immunologic tolerance might be operative commonly in preventing host rejection of tumor. Hence, development of methods aimed at breaking immunologic tolerance for treating cancer should be worthy of major attention. Freund's adjuvant, especially when given with added Mycobacteria, is highly effective in overcoming immunologic nonresponsiveness even to the point of promoting immunologic reaction to the host's own normal tissues (44). It has also recently been shown that immunologic paralysis to pneumococcal polysaccharide may be broken by administration of endotoxin (45). Sparck et al. (46) presented evidence to indicate possible breakage of tolerance to mouse leukemia virus by adoptive transfer of competent donor cells.

The essential role of thymus for development of immunologic competence is well established. The thymus appears to be the initial source of lymphoid cells which become peripheralized to the spleen and lymph nodes and additionally, thymus or thymus hormone action is essential to maturation of lymphoid cells. Thymectomy of newborn animals greatly enhances their capacity to develop virus-caused cancer, even in species or strains that are normally resistant (47–49). This emphasizes the need to study the relationship of thymus to cancer and emphasizes the desirability of attempting to enhance or to restore immunologic responsiveness in an immunologically laggard tumor-bearing host. Additionally, stimulation of thymic activity to promote development of immunologic competence at the earliest age might be of advantage in preventing hypothetical virus-induced neoplastic transformation in the newborn. Sherman and others (50) have been able to induce apparent thymic stimulation in animals by injecting any of a variety of antigens directly into the thymus. Plasma cells and follicles developed in the thymus and an anemia appeared which was possibly of autoimmune type. Such procedure, if not carried to the point of loss of self-recognition of normal antigens, might be of benefit in stimulating host reactivity to cancer.

Interferons are viral inhibitory substances of protein composition that are commonly formed by cells in response to stimulation by viruses and by other microbes or their components (51–54). Interferons commonly appear in the blood and tissues of animals and man in the course of viral and other infections and appear to provide a mechanism for recovery from viral disease which is separate and distinct from specific immunologic mechanisms. Exogenous interferons, administered prophylactically or therapeutically, presently show little promise in clinical medicine for preventing or treating viral disease. On the contrary, stimulation of endogenous interferon in the host itself by appropriate substances may contribute significantly to enhancement of host defenses. Interferon may be a mechanism whereby the lytic process of virus in cells is limited thereby permitting a state of persistent infection, and any consideration for control of cancer caused by viruses must necessarily take account of the importance of interferon in viral infection. Principal hope for utilization of the interferon mechanism in control of acute infectious disease caused by viruses lies in the effort to stimulate the body to produce its own interferon, using a nontoxic substance. Present evidence indicates that viruses, foreign nucleic acids, certain bacteria, coli lipopolysaccharide, and polysaccharides such as statolon, all may stimulate production or release of interferons. Any procedure which is practical in preventing or limiting infection with viruses should be of use in preventing cancer caused by viruses. Alternatively, it might be of equal importance to develop methods for negating interferon action, possibly permitting latent virus or integrated viral genome in neoplastic cells to gain control causing lysis and eliminating the undesirable cells.

A relationship of interferon to neoplasia has been demonstrated in a few examples to date: Studies in our laboratories (55) revealed marked suppression in number and size of sarcomas in chicks infected with Rous sarcoma virus when purified and concentrated chick interferon was given prophylactically. Friedman and Rabson (56) studied a highly oncogenic and a weakly oncogenic strain of polyoma virus in mice. Extracts of tissues of mice infected at birth with the poorly oncogenic variant revealed the presence of interferon whereas tissues of mice infected with the highly oncogenic strain contained no detectable interferon. Todaro and Baron (57) recently showed that interferon was capable of preventing malignant transformation by SV40 virus of mouse cell line 3T3 in the absence of replication of the virus. The findings support the conclusion that interferon may act intracellularly to block an event which is essential to neoplastic transformation.

Closing Remarks

It may be concluded that in spite of the promise which the immunologic approach to control of cancer holds, the more than half century of investigation of specific and nonspecific immunologic approaches to cancer have failed to yield one procedure of significant application to man. The principal problems of cancer still remain: Seek the cause and seek the cure. If the cause can be found and can be prevented, the cure will not be necessary.

References

1. EPSTEIN, M. A., WOODALL, J. P., and THOMSON, A. D., 1964: Lymphoblastic lymphoma in bone-marrow of African green monkeys (Cercopithecus Aethiops) inoculated with biopsy material from a child with Burkitt's lymphoma. Lancet 2, 288–291.
2. MCALLISTER, R. M., LANDING, B. H., and GOODHEART, C. R., 1964: Isolation of adenoviruses from neoplastic and non-neoplastic tissues of children. Lab. Invest. 13, 894–901.
3. BELL, T. M., MASSIE, A., ROSS, M. G. R., and WILLIAMS, M. C., 1964: Isolation of a reovirus from a case of Burkitt's lymphoma. Brit. Med. J. 1, 1212–1213.
4. EPSTEIN, M. A., ACHONG, B. G., and BARR, Y. M., 1964: Virus particles in cultured lymphoblasts from Burkitt's lymphoma. Lancet 1, 702–703.
5. DALLDORF, G., and BERGAMINI, F., 1964: Unidentified, filtrable agents isolated from African children with malignant lymphomas. Proc. Nat. Acad. Sci. 51, 263–265.
6. BURGER, C. L., HARRIS, W. W., ANDERSON, N. G., BARTLETT, T. W., and KNISELEY, R. M., 1964: Virus-like particles in human leukemic plasma. Proc. Soc. Exp. Biol. Med. 115, 151–156.
7. SMITH, K. O., BENYESH-MELNICK, M., and FERNBACH, D. J., 1964: Studies on human leukemia. II. Structure and quantitation of myxovirus-like particles associated with human leukemia. J. Nat. Cancer Inst. 33, 557–570.

8. DMOCHOWSKI, L., 1965: Electron microscopic observations of leukemia in animals and in man. Cancer Res. **25**, 1654–1671.

9. EPSTEIN, M. A., BARR, Y. M., and ACHONG, B. G., 1965: Studies with Burkitt's lymphoma. Methodological Approaches to the Study of Leukemias, V. Defendi (Ed.). Wistar Institute Symposium Monograph No. 4, 69–79.

10. STEWART, S. E., LANDON, J., LOVELACE, E., and PARKER, G., 1965: Burkitt tumor: Brain lesions in hamsters induced with an extract from the SL-1 cell line. Methodological Approaches to the Study of Leukemias, V. Defendi (Ed.). Wistar Institute Symposium Monograph No. 4, 93–101.

11. HENLE, G., and HENLE, W., 1965: Interference in the detection of viral carrier states. Methodological Approaches to the Study of Leukemias, V. Defendi (Ed.). Wistar Institute Symposium Monograph No. 4, 83–90.

12. O'CONOR, G. T., and RABSON, A. S., 1965: Herpes-like particles in an American lymphoma: Preliminary note. J. Nat. Cancer Inst. **35**, 899–903.

13. ANDERSON, D. R., 1965: Subcellular particles associated with human leukemia as seen with the electron microscope. Methodological Approaches to the Study of Leukemias, V. Defendi (Ed.). Wistar Institute Symposium Monograph No. 4, 113–141.

14. MURPHY, W. H., ERTEL, I. J., and ZARAFONETIS, C. J. D., 1965: Virus studies of human leukemia. Cancer **18**, 1329–1344.

15. GRACE, J. T., JR., HOROSZEWICZ, J. S., STIM, T. B., MIRAND, E. A., and JAMES, C., 1965: Mycoplasmas (PPLO) and human leukemia and lymphoma. Cancer **18**, 1369–1376.

16. ARMSTRONG, D., HENLE, G., SOMERSON, N. L., and HAYFLICK, L., 1965: Cytopathogenic mycoplasmas associated with two human tumors. I. Isolation and biological aspects. J. Bact. **90**, 418–424.

17. HUMMELER, K., TOMASSINI, N., and HAYFLICK, L., 1965: Ultrastructure of a mycoplasma (Negroni) isolated from human leukemia. J. Bact. **90**, 517–523.

18. HAYFLICK, L., 1965: Mycoplasmas and human leukemia. Methodological Approaches to the Study of Leukemias, V. Defendi (Ed.). Wistar Institute Symposium Monograph No. 4, 157–164.

19. DMOCHOWSKI, L., TAYLOR, H. G., GREY, C. E., DREYER, D. A., SYKES, J. A., LANGFORD, P. L., ROGERS, T., SHULLENBERGER, C. C., and HOWE, C. D., 1965: Viruses and mycoplasma (PPLO) in human leukemia. Cancer **18**, 1345–1368.

20. BURMESTER, B. R., WALTER, W. G., and FONTES, A. F., 1957: The immunological response of chickens after treatment with several vaccines of visceral lymphomatosis. Poultry Sci. **36**, 79–87.

21. FINK, M. A., and RAUSCHER, F. J., 1961: A simple method for the preparation of potent chicken anti-Rous sarcoma virus serum. J. Nat. Cancer Inst. **26**, 519–522.

22. HUGHES, W. F., WATANABE, D. H., and RUBIN, H., 1963: The development of a chicken flock apparently free of leukosis virus. Avian Dis. **7**, 154–165.

23. JENSEN, F. C., GIRARDI, A. J., GILDEN, R. V., and KOPROWSKI, H., 1964: Infection of human and simian tissue cultures with Rous sarcoma virus. Proc. Nat. Acad. Sci. **52**, 53–59.

24. FRIEND, C., 1959: Immunological relationships of a filterable agent causing a leukemia in adult mice. I. The neutralization of infectivity by specific antiserum. J. Exp. Med. **109**, 217–228.

25. FINK, M. A., and RAUSCHER, F. J., 1964: Immune reactions to a murine leukemia virus. I. Induction of immunity to infection with virus in the natural host. J. Nat. Cancer Inst. **32**, 1075–1082.

26. FINK, M. A., RAUSCHER, F. J., and CHIRIGOS, M., 1965: Some immune reactions of murine leukemia viruses demonstrable within a completely isologous system. Presented at Workshop on Prospects for Control of Viral-Induced Tumors by Immunologic Methods and Chemotherapy, held by Subcommittee on Carcinogenesis and Prevention, National Advisory Cancer Council. June 18–19, 1965, Bethesda, Maryland. J. Nat. Cancer Inst. *To be published*.

27. SINKOVICS, J. G., BERTIN, B. A., and HOWE, C. D., 1965: Some properties of the photodynamically inactivated Rauscher mouse leukemia virus. Cancer Res. **25**, 624–627.

28. EDDY, B., STEWART, S. E., and TOUCHETTE, R., 1959: Effect of immunization of adult female hamsters on the latency of infection in offspring inoculated with the SE polyoma virus. Fed. Proc. **18**, 565.

29. EDDY, B. E., GRUBBS, G. E., and YOUNG, R. D., 1964: Tumor immunity in hamsters infected with adenovirus type 12 or simian virus 40. Proc. Soc. Exp. Biol. and Med. **117**, 575–579.

30. DEICHMAN, G. I., and KLUCHAREVA, T. E., 1964: Prevention of tumour induction in SV40 infected hamsters. Nature **202**, 1126–1128.

31. SAITO, J. K., McKERCHER, D. G., and CASTRUCCI, G., 1964: Attenuation of the myxoma virus and use of the living attenuated virus as an immunizing agent for myxomatosis. J. Infect. Dis. **114**, 417–428.

32. HUEBNER, R. J., LANE, W. T., WELCH, A. D., CALABRESI, P., McCOLLUM, R. W., and PRUSOFF, W. H., 1963: Inhibition by 5-iododeoxyuridine of the oncogenic effects of adenovirus type 12 in hamsters. Science **142**, 488–490.

33. FISCHER, D. S., BLACK, F. L., and WELCH, A. D., 1965: Inhibition by nucleoside analogues of tumour formation by polyoma virus. Nature **206**, 839–840.

34. GRAHAM, J. B., and GRAHAM, R. M., 1962: Autogenous vaccine in cancer patients. Surg. Gyn. Obst. **114**, 1–4.

35. DONIACH, I., CROOKSTON, J. H., and COPE, T. I., 1958: Attempted treatment of a patient with choriocarcinoma by immunization with her husband's cells. J. Obst. Gyn. (Brit. Comm.) **65**, 553–556.

36. HACKETT, E., and BEECH, M., 1961: Immunological treatment of a case of choriocarcinoma. Brit. Med. J. **2**, 1123–1126.

37. CINADER, B., HAYLEY, M. A., RIDER, W. D., and WARWICK, O. H., 1961: Immunotherapy of a patient with choriocarcinoma. Canad. Med. Assoc. J. **84**, 306–309.

38. GOLDNER, H., GIRARDI, A. J., and HILLEMAN, M. R., 1963: Attempts to interrupt virus tumorigenesis by immunization using homologous "Bjorklund-type" antigen. Proc. Soc. Exp. Biol. and Med. **114**, 456–467.

39. GOLDNER, H., GIRARDI, A. J., LARSON, V. M., and HILLEMAN, M. R., 1964: Interruption of SV40 virus tumorigenesis using irradiated homologous tumor antigen. Proc. Soc. Exp. Biol. and Med. **117**, 851–857.

40. GOLDNER, H., GIRARDI, A. J., and HILLEMAN, M. R., 1965: Enhancement of virus tumorigenesis in hamsters attending vaccination procedures. Virology **27**, 225–227.

41. GIRARDI, A. J., 1965: Prevention of SV40 virus oncogenesis in hamsters, I. Tumor resistance induced by human cells transformed by SV40. Proc. Nat. Acad. Sci. **54**, 445–451.

42. WOODRUFF, M. F. A., 1964: Immunological aspects of cancer. Lancet **2**, 265–270.

43. MATHE, G., AMIEL, J. L., SCHWARZENBERG, L., CATTAN, A., and SCHNEIDER, M., 1965: Adoptive immunotherapy of acute leukemia: Experimental and clinical results. Symposium: Conference on Obstacles to the Control of Acute Leukemia, Warrenton, Va., March 21–23, 1965. Cancer Res. **25**, 1525–1531.

44. Waksman, B. H., 1962: Auto-immunization and the lesions of auto-immunity. Medicine 41, 93–141.

45. Brooke, M. S., 1965: Conversion of immunological paralysis to immunity by endotoxin. Nature 206, 635–636.

46. Sparck, J. V., and Volkert, M., 1965: Effect of adoptive immunity on experimentally induced leukaemia in mice. Nature 206, 578–579.

47. Defendi, V., and Roosa, R. A., 1965: Effect of thymectomy on induction of tumors and on the transplantability of polyoma-induced tumors. Cancer Res. 25, 300–306.

48. Ting, R. C., and Law, L. W., 1965: The role of thymus in transplantation resistance induced by polyoma virus. J. Nat. Cancer Inst. 34, 521–527.

49. Kirschstein, R. L., Rabson, A. S., and Peters, E. A., 1964: Oncogenic activity of adenovirus 12 in thymectomized BALB/3 and C3H/HeN mice. Proc. Soc. Exp. Biol. and Med. 117, 198–200.

50. Sherman, J. D., Adner, M. M., and Dameshek, W., 1964: Direct injection of the thymus with antigenic substances. Proc. Soc. Exp. Biol. and Med. 115, 866–870.

51. Wagner, R. R., 1965: Interferon. A review and analysis of recent observations. Amer. J. Med. 38, 726–737.

52. Ho, M., 1964: Identification and "induction" of interferon. Bact. Rev. 28, 367–381.

53. Hilleman, M. R., 1963: Interferon in prospect and perspective. J. Cellular Comp. Physiol. 62, 337–353.

54. Hilleman, M. R., 1965: Immunologic, chemotherapeutic and interferon approaches to control of viral disease. Amer. J. Med. 38, 751–766.

55. Lampson, G. P., Tytell, A. A., Nemes, M. M., and Hilleman, M. R., 1963: Purification and characterization of chick embryo interferon. Proc. Soc. Exp. Biol. and Med. 112, 468–478.

56. Friedman, R. M., and Rabson, A. S., 1964: Possible role of interferon in determining the oncogenic effect of polyoma virus variants. J. Exp. Med. 119, 71–81.

57. Todaro, G. J., and Baron, S., 1965: The role of interferon in the inhibition of SV40 transformation of mouse cell line 3T3. Proc. Nat. Acad. Sci., U.S. 54, 752–756.

Round Table Discussion

Dr. Huebner, *Chairman*

DR. HUEBNER: After all this discussion I am wondering about the audience. I think it would help, if someone would make some suggestions about how one persuades these investigators, when they talk about the same things, to use the same words. I think this would help the audience. If I was sitting in the audience right now, I would applaud that suggestion! It is hard enough to follow some of the very involved experiments that have been described today, without having to figure out that SP2 means E46 and T and neoantigen something else. We do need a glossary of terms; we also need a lexicon of defined words that describe what we are talking about. I believe that the geneticists have arrived at some sort of agreement on this point; I think the cancer biologists with their new antigens might also get together and decide to agree on common terms. I have proposed a lexicon for the T-antigens and the antigens found in hamster tumors (Table I).

I did want to make a couple of points that have not been made in this meeting; we have not talked about all possible oncogenic systems. I just wanted to mention in passing that the number of oncogenic adenoviruses is not limited to those discussed here. It is now known that eight of the thirty-one human adenoviruses are oncogenic, that six of the eighteen simian adenoviruses are oncogenic and that one of three bovine adenoviruses and one of two avian adenoviruses produce tumors in hamsters. The oncogenic property of adenoviruses, of course, refers to a group of agents that occurs in many different species. I think this in itself is important. One of the things about the adenovirus oncogenesis that has not been mentioned here today is that the determinants of oncogenesis in the case of adenoviruses are closely associated with other determinants that result in the cellular T-antigens. They are also associated with the molecular weight and the base composition of the DNAs of the adenoviruses, and also with those responsible for hemagglutinins.

If one categorizes adenovirus oncogenesis for hamsters into a) highly oncogenic, b) moderately oncogenic, and c) non-oncogenic, one finds that the adenoviruses distribute almost exactly the same way with respect to T-antigens. The T-antigens of the highly oncogenic group are shared, amongst each other and not with others; the T-antigens of the moderately oncogenic group, such as types 3, 7, 14, 16, and 21, and perhaps also 11, also share a common T-antigen. Those that do not produce tumors do not have T-antigens which are shared with the others, with one exception. In addition, the G + C ratios of adenovirus DNAs described by Pina and Green distribute into groups according to oncogenic activity. The lowest G + C ratio, 47 to 49, turns out to belong in the a) group. The b) group, which is moderately oncogenic, has a slightly heavier DNA and a G + C content between 51 and 53. Adenoviruses that do not cause tumors, and this includes most of them,

TABLE I

LEXICON OF REAGENTS

---

1. *Preparations of Antigen*

T Antigen (Neoantigen) Cell Pack

Suspensions of infected cells (usually 10 to 20 per cent packed cells in fluid media) collected before CPE (most often 24, 48, or 72 hours after inoculation) clarified at 1500 revolutions per minute for 15 minutes. T antigens are newly recognized virus-specific but nonvirion antigens that are detectable with specific antisera from hamsters or other animals carrying virus-induced tumors. Unless blocked by cytosine arabinoside or FUDR, these preparations also contain virion antigens.

Viral (V) Antigen

Unconcentrated virus-infected cells and fluids homogenized in a conventional way and harvested at optimum time for maximum infectivity, clarified at 1500 revolutions per minute for 15 minutes. They contain mostly virion but may also contain T antigens in variable amounts.

Virion

Concentrated and purified viral particles; usually infectious; separated from cellular materials and cellular antigens by gradient centrifugation or by other means.

Virus Subunit

Purified preparations of subunits of the virion, such as A, B, and C antigens of adenovirus (obtained by chromatography) and S and V antigens of myxoviruses.

Viral Cell Pack

Packed cells in tissue culture are taken after CPE has become widespread to give maximum infectious virion antigen; the cells are homogenized in supernatant media to give a 20 per cent suspension; they may contain some T antigens (cell-associated or neoantigens) as well.

Tumor Antigen

Homogenized, 20 per cent suspension of virus-induced tumor cells from the hamster or other animal; clarified at 1500 revolutions per minute for 15 minutes.

Tumor-Cell Culture

Ten to twenty per cent cell-pack suspension of tumor cells grown *in vitro* in tissue cultures.

Isoantigen

Preparations of cells in tissue culture, spontaneous tumors, and normal tissues from hamster, rat, or other species which react nonspecifically in complement fixation (CF) or fluorescent-antibody tests with the sera from various mammalian species. These antigens must be recognized insofar as possible whenever present in specific tumor-virus systems.

2. *Preparations of Antiserum*

Antivirion Antiserum

Specific antisera usually resulting from acute viral infection or immunization with purified virions. Contains antibodies directed against the virus particle; they usually neutralize the virus, for example, polyoma virus in the mouse, SV40 in the monkey, adenovirus in man. Virus-specific antibodies to either tumor antigens or T antigens are usually not present.

TABLE I (*continued*)

Antitumor Antiserum	Sera from animals carrying a virus-induced tumor. Reaction is mainly with homologous tumor extracts, cultured tumor-cell antigens, and T antigen cell packs. In some cases, they may also contain antivirion and neutralizing antibodies (as in adenovirus type 12). There are several types: A. Narrow-reacting. (React only with tumor and T antigen cell-pack preparations.) B. Broad-reacting. (React with viral (V) conventional preparations, as in adenovirus type 12.) This reaction needs more precise definition; it may reflect only stronger reactions with the small amounts of T antigens in viral preparations. C. Broadest-reacting. (Sera of hamsters, rats, or other tumor-bearing animals which react specifically with virion or viral-subunit antigens, as with C antigen of adenovirus type 12; it may neutralize virus.

are the heaviest, showing a G + C ratio of 53 to 57. So, we have two determinants that are correlated very closely with the oncogenic properties of the adenoviruses. The third one is the adenovirus hemagglutinins (HA). The adenovirus HA were described by Rosen (Rosen, L.: *Amer. J. Hyg.* 71: 120–128, 1960), who classified them into four groups (Table II). The highly oncogenic group a) is classified in group 4) and makes up the entire group. The moderately oncogenic group b) belongs in HA group 1). Those adenoviruses that are not oncogenic belong primarily in groups 2) and 3)—almost entirely in 2). We have now three properties of adenoviruses that are closely associated with the determinants of oncogenesis, and this, of course, provides us with different markers for studying the properties of oncogenic adenoviruses as compared to those which are not.

One other point that has not been mentioned today is the newly described mouse sarcoma virus (MSV) described by Harvey and by Moloney. They were both derived from the Moloney leukemia and they produce foci in tissue culture, as described by Drs. Hartley and Rowe in my lab. At the same time, MSV produces a virus that is transmissible, and appears to be like the Rous sarcoma virus in that it is defective, in the sense that in order to produce focal transformation in mouse embryo fibroblasts (MEF), it needs a helper virus. The helper can be any one of a number of different mouse leukemia viruses. This is about as far as we can go at this point in pointing out the similarity to the avian leukosis viruses. We do not yet know about the viral antigens, or the internal antigens, except that there is as in the avian tumor viruses also a common CF antigen in the mouse leukemias. We are not sure whether this common CF antigen is an internal antigen, or an envelope antigen. Finally, the gist of what I have said is this: There are patterns that appear to be consistently expressed by the tumor viruses, and these patterns are coming more and more to the fore all the time. I am sure that the various markers and the various parameters we now have for studying these viruses in *in vitro* systems are

TABLE II

ADENOVIRUS GROUPINGS ACCORDING TO HA CHARACTERISTICS (ROSEN *)

HA Groups	Serotypes Included in HA Groups	HA Erythro-cytes of Rhesus	Rats	Demonstrated Oncogenesis	DNA Base Composition % G–C Ratio	T Antigen in CF
1	3, 7, 11, 14, 16, 20, 21, 25, 28	+	0	Intermediate 3, 7, 14, 16, 21	Intermediate (50–53) 3, 7, 11, 14, 16, 21	3, 4, 7, 11, 14, 16, 21 Positive for Group B
2	8, 10, 13, 15, 17, 19, 22–24, 26, 27, 29, 30	+	+	Not Demonstrated	High (56–60) Includes 20, 25, 28 from HA Group 1	Negative for Group A or B
3	1, 2, 4–6	0	Partial	Not Oncogenic	High (56–58)	Negative for Group A or B
4	12, 18, 31	0	±	Highly Oncogenic	Low (47–49) 12, 18, 31	12, 18, 31 Positive for Group A

* Rosen, L.: *Amer. J. Hyg.* **71**:120–128, 1960.

going to help make it possible to determine the mechanisms of viral oncogenesis.

We have some questions from the audience. First a question for Dr. Habel: Most human cancers occur in the adult or aging population. Can one infer that a) the cancer started in the immunologically immature neonate or b) the cancer develops because the older individual becomes immunologically less competent?

DR. HABEL: Anyone who develops a theory based on experimental data can always find a justification for this in other known medical facts. With the results of the polyoma and SV40 studies concerning the transplantation antigens and the effects of immunological competence on the development of tumors, it is interesting to note that the two extremes of life, of human life, are the times of increased incidence of tumors. The two extremes of age are also the extremes of immunological competence. At least, the evidence suggests that. If one wants to be rather broad in the application of experimental data, one could say that this could be of some significance. So far, there is no evidence that a tumor can be suppressed immunologically over a period of time in an experimental system by immunological rejection, and then at some later date, due to a lowering of immunological competence, a persisting transformed cell that had been held under control immunologically suddenly starts to divide. Some attempts have been made to try to demonstrate this, but as far as I know, they have not been successful.

DR. HUEBNER: Here is another general question, and I suppose almost anyone here could say something about it: "Please discuss the biochemical changes in malignant cell transformation—DNA, protein deletion, and h_2 protein, etc. I think perhaps the DNA and protein deletion might be particularly worth discussing."

DR. RAPP: The biochemical events during transformation, in many of the DNA virus systems, are difficult to study because it takes so long for morphologic transformants to appear after exposure to virus. The other approach taken by a number of investigators has been to measure the change in the synthesis of DNA or in level of enzymes following introduction of some of the oncogenic viruses during the early phases of the cytolytic cycle. One of the striking changes that has been observed is that, unlike other DNA viruses that are not oncogenic and which cause the shutdown of the cell DNA biosynthetic capacity, viruses such as SV40 and polyoma actually cause an increase in the synthesis of DNA. One can measure this, for instance, by uptake of tritiated thymidine, and if one measures the enzymes involved in DNA synthesis, such as thymidine kinase, one finds increased levels following infection. The properties of some of these enzymes appear to be different from host cell enzymes; that is, they may have different Michaelis constants, different heat stabilities, or differ antigenically. I know of no one who has searched for these enzymes in transformed cells. However, adequate controls for transformed cells are very hard to come by, since many of the transformed cells have been maintained for quite some time in *in vitro* passages.

DR. HUEBNER: Here is a question from the chairman, Dr. Roizman: Has anyone looked for reciprocal SV40 adenovirus hybrids, that is, adeno DNA with an SV40 coat? Would someone predict the properties of such a hybrid?

DR. RAPP: No, we have not looked for it. One of the difficulties would be that the SV40 is considerably smaller than the adenovirus. In addition, the adenovirus contains much more DNA than the SV40 virus. It would be a little harder, I would guess, to pack the adeno-DNA into an SV40 capsid than it is to put some SV40 DNA into an adenocapsid. We have made preliminary attempts at doing so, but have been unsuccessful. What we have tried to do, however, in that connection, is to see if we can put some herpes virus protein around the SV40 DNA but we have not succeeded in accomplishing this either.

DR. HUEBNER: Here is another question from the audience: "Which cells in a mouse embryo culture respond to polyoma virus by vegetative response, and which cells eventually undergo transformation? Why do rat and hamster cells transform best and produce little if any infectious virus? Is interferon involved here?"

DR. GIRARDI: Studies on interferon production by cells which are transformed by viruses seem to indicate that even where interference occurs, it is mediated through a system other than the production of interferon. One can go a little further: physical, chemical and viral oncogenic agents in a system where transformation occurs usually do not produce interferon. This could very well be an important step, because if the interferon is involved in the control of RNA synthesis in a cell, it is one property that appears to be deleted in those situations. Even in the Rous sarcoma virus situation, the interferon does not seem to be formed, but the interference occurs at a different level. In the SV40-human cell system, where interference with polio-virus synthesis occurs in some of the transformed cells, it is not mediated through interferon. With chemical and physical agents, the work of the De Maeyers has shown these agents suppress the formation of interferon.

As far as the question about infectious virus and non-infectious virus is concerned, it would seem that those cells that are capable of overcoming the effects of lytic infection survive. Cells with an abortive cycle should go on to become transformed. I have some data I'll present tomorrow concerning this in the human cell.

DR. MACPHERSON: A comment on the relative sensitivities of cells to lysis and transformation. In hamster and rat cells there is very little cell lysis, if any at all. One can do accurate quantitative assays on the respective sensitivities of cells to transformation. If one takes primary hamster embryo cells as a sensitivity of "one" as judged by transformation assays, the BHK-21 cell would be something like 300 times more sensitive in platings or in the agar suspension assay. Rat fibroblasts are approximately the same; the rat cell is much more susceptible than the hamster in this respect.

DR. HUEBNER: Here is a question which, I suppose, is aimed at Dr. Temin: What is the evidence at this time that malignant conversion of cells infected with RNA tumor viruses, such as Rous sarcoma virus, is based on incorporation into the host cell genome of DNA that is complementary to viral RNA?

DR. TEMIN: The hypothesis that malignant conversion of cells infected with Rous sarcoma virus is based on incorporation into the host cell genome of DNA that is complementary to viral RNA, or more simply that the provirus is DNA, rests on three types of evidence. First, experiments using actinomycin D, an antibiotic which stops DNA-directed RNA synthesis, showed that virus production and viral RNA synthesis in already virus producing cells was blocked by this agent. Second, experiments with other antibiotics established a requirement for new DNA synthesis at the time of infection. Third, hybridization experiments with labeled RNA from purified virus showed an increased amount of homology with DNA from infected as compared to uninfected cells. We have now started to do these hybridizations with P-32 labeled virus. Attempts to demonstrate directly what could be called the preprovirus have so far failed due to technical problems. One can postulate that after infection the viral RNA would make a DNA which would be found in the cytoplasm and this could be demonstrated.

DR. HUEBNER: Now, the question here has two parts, and the first one had to do with malignant conversion rather than viral replication. You were talking more about viral replication in your answer. Do you have anything more to say with respect to conversion?

DR. TEMIN: Since it has been shown that infection of cells and formation of a provirus is necessary for both virus replication and conversion, we can study either to find out the molecular nature of the provirus.

The evidence which I was referring to in the experiments with the antibiotics involved virus production. As far as the hybridization experiments are concerned, these were with DNA from converted cells. We could say that in these cells there was more DNA homologous to viral RNA than there was in uninfected cells.

DR. HUEBNER: Since it appears to be fairly evident that the non-virus producing cell does contain viral protein, possibly from the internal component, one would not even have to postulate incorporation of the viral genome in the cell genome.

DR. TEMIN: That is true. The necessity for the provirus was originally based on genetic evidence of the control of conversion in the infected cells, independent of the presence or absence of virus protein. One must ask how the information controlling conversion and virus production is transmitted as the cells divide. The original work showed that there was something in the cells which was regularly inherited, and which only had one or two copies, the provirus. This was based on genetic evidence, independent of the existence of abortive cycles of virus replication. Then, the question arose what is the molecular nature of this provirus? It could have been double-stranded RNA or DNA. The evidence suggests that it is DNA. Further, the question arises as to the activity of this DNA and its function in causing conversion and causing virus production. It appears that it must make an RNA which is present in the virus, something about 10^7 MWU per particle; it must make RNA which would code for virus protein and make virus. Since it is possible to have cells which are producing virus and are not converted, and cells which are converted and not producing infectious virus, it also must code in another way for things which bring about conversion.

DR. SABIN: I want to ask Dr. Temin why he calls viral genetic material a provirus, which is capable of synthesizing viral material and viral antigens. This is not quite the definition of provirus that is used in college. Could you elucidate?

DR. TEMIN: In lysogeny the term that is used is *prophage*. In infection of bacteria with a temperate phage there are two possible cycles: one is the lytic cycle, in which the viral genome seems to replicate vegetatively and the cell lyses with production of progeny virus. There is another cycle in which the genetic information of the virus is incorporated in one or two copies within the cell genome and is regularly inherited. This regularly inherited structure is called the prophage. Now as I just indicated, the genetic evidence from experiments which we carried out shows that in the cells infected with the Rous sarcoma virus there are one or two copies of a regularly inherited structure which has information for the production of the virus. Therefore, this is called, by analogy, a provirus. Then, the question arises is the provirus the same molecule as the RNA in the virus, or is it something else like replicative form, or is it DNA, as I postulated?

DR. HUEBNER: Well the question I would like to have you tackle is this: first of all, how do you distinguish between defective virus and provirus, and second, how is the non-producer cell different from a carrier cell that is carrying an incomplete virus particle?

DR. TEMIN: Defective virus is something applying to an entire life cycle; provirus is a term for one part of the cycle. We would assume that in a life cycle of a Rous virus it would start out with the infectious virus, the virion, containing RNA, protein and lipid. After infection of the cell, this would go into some state which we call pre-provirus; you could call it replicative form, perhaps molecularly this is an RNA-DNA hybrid. After this there is something we call a provirus, which we have postulated is double-stranded DNA integrated with the cell genome. This, we assume, makes a messenger RNA, and also an RNA which would be the RNA of the virus, and this would come out of the nucleus and at the cell membrane make the virion. The infecting virus, pre-provirus, provirus, and virion are all stages in a circular life cycle. You can have different types of life cycles: the abortive infection by a defective virus, or complete infection by virus. This would be determined by whether or not the genetic information in the single molecule of nucleic acid in the virion was enough to carry information for the entire cycle or not. So, defective virus refers to the whole cycle; provirus to a stage in the cycle. There can, of course, be defective proviruses.

The molecular basis of the converted non-infectious virus producing cell is not clear. There are several hypotheses about this. The one which I prefer is the one from analogy with lysogeny. Somehow, in this cell that has the genetic information, the provirus, some part of the messenger for virus production is not expressed, and therefore there is only partial synthesis of the virion. Another possibility, which is described by the term defective, is that the provirus actually is missing the nucleotide sequences which would specify some part of the virus. This defect was originally thought to be in information for the synthesis of the virus coat, but the work of Dougherty and Di Stefano suggests that it is not this, but something else. The two hypotheses differ in the nature of what information is present in the provirus. In other words, we have a sequence of nucleotides in the provirus: one hypothesis states these nucleotides are sufficient to code for all of the virion, the other that they are not sufficient.

DR. HUEBNER: The thing I tried to get clear here was the uniqueness of Rous sarcoma provirus which I failed to understand and still cannot quite accept because there have been reports—by Henle and others—who have produced carrier cell cultures of mumps virus and other myxoviruses. In this case, they have a whole virus particle present except the envelope, and Vogt is postulating now that most avian sarcoma non-producer cells have everything of the virus particle except the envelope and that something else has to supply that.

Dr. Temin: The type of thing you are talking about was first described actually by Walker and Hinze in mumps infection of conjunctiva cells. This type of system they have called a "regulated culture." When you look at a population of cells and you find the cell number is increasing and there is also virus produced, there are several possible explanations. One is a carrier culture in which only some of the cells in the culture are infected and produce virus in which case they die. Others are not infected—we are not talking about this. Then, at the time we started this work, the only other system was the Rous sarcoma virus system where no cells died and all cells were producing virus. It was then shown in 1961 in the Rous case that a provirus existed. Now, since about 1962, it has been obvious with especially the myxovirus II-group that there is some kind of state possible in which virus can exist, probably in the cytoplasm, and replicate by an RNA-RNA mechanism without killing the cell. The killing of the cells is separable. Therefore, the population studies which originally separated the type of state of the Rous virus from the type of state of the carrier culture would not separate the regulated culture from the Rous or integrated state. However, there are still distinctions which exist between the regulated culture and the integrated (Rous) culture. In the regulated culture there has been no sign of the one or two regularly inherited units. In fact, it has been possible in Walker's case to cure these cultures by changing the temperature and changing the conditions, in which case the cells will die or will be cured. In the case of the Rous, by double infection experiments it was shown that there had to be some direct interaction with the host so the information in the provirus was regularly inherited independent of environmental conditions. The explanation we would give at the present time of the regulated culture is this: these viruses are replicating by typical RNA-double-stranded mechanisms (RF) in the cytoplasm; they are somehow deficient in what is necessary to kill a cell. Therefore, as a cell divided, these RF are given to the daughter cells by a process similar to that for any cytoplasmic particles. In the Rous case, we believe there is an RNA to DNA replication, so that integrated DNA, provirus, is given to the daughter cells equally at division.

Dr. Huebner: I think, perhaps, just as a postscript to that, that I might quote Dr. Rowe from a couple of weeks ago at the "Perspectives in Virology" meeting in New York. I think many other people agree with this. A virus particle that is oncogenic is defective in the cell that it is going to transform. Perhaps in nearly every case some helper virus may be needed to maintain the infectious virus particle in nature.

Dr. Temin: Certainly with the leukemia viruses and some of the Rous strains this would not be true. However, it appears to be possible in the Rous case for a converted cell not to produce infectious virus, and in this sense for the infection to be abortive. But, even though a provirus is formed it is certainly not a necessary condition. With the Schmidt-Ruppin virus, the virus can replicate in the tumor cell. By superinfecting a tumor cell or by mixedly infecting a cell with virus from the Bryan high-titer strain and a leukosis virus, one can get all cells in Rous sarcomas to produce virus, without affecting their being tumor cells. In the Rous sarcoma virus-chicken system there is no necessity not to produce infectious virus. I think it is proper to say that the production of infectious virus is not a necessary part of oncogenesis, but that it can occur or it cannot. What is necessary is the formation of a provirus.

DR. GIRARDI: When it does occur in the Schmidt-Ruppin Rous sarcoma virus system, what is the percentage of cells that yield infectious virus?

DR. TEMIN: It is nearly a hundred percent, although we have a few cases of converted non-infectious virus producing clones.

DR. RAPP: I think Dr. Temin raised a point that, at the moment at least, differentiates the Rous virus system from the DNA-transforming viruses. While one can rescue Rous sarcoma virus from transformed cells by superinfecting the cells with any member of the avian leukosis complex, no one so far has been able to do this with cells transformed by a DNA virus. Rescue of the oncogenic determinants has not been possible. These cells are quite often very resistant to superinfection with the same virus, which the Rous sarcoma cells do not appear to be.

DR. HUEBNER: Thank you, Dr. Rapp. Here is another question from the audience: Poel has advanced the idea that viral, genetic or carcinogenic factors lead to "incapacitated cells," and that normal cells, minus the inhibitory influence of the incapacitated cells, lead to tumor development and progression. Could someone comment on this thought? We would have to define incapacitated cells, somehow.

DR. HABEL: I gather that this raises the question as to whether or not it is *the* cell which is invaded and which has an intimate interaction with the virus that is transformed, or whether the infected cell has some indirect influence on the surrounding cells and they are the ones that are transformed. Well, I think that the *in vitro* studies of Dr. Macpherson and others, showing the direct relationship of the dose of virus to the number of transformed cells, the fact that the transformed cells are the ones that have the specific antigens in them, all this, certainly, is against any indirect mechanism, and I think has established for the first time in recent years. It is the very cell which is infected and with which the virus makes an effective contact, that is transformed.

DR. DAWE: I would like to comment on Dr. Habel's remarks with reference to our *in vitro* system of organ cultures. At one stage in our work we thought that perhaps the function of polyoma virus was to "kill off" the mesenchymal cells and liberate them, theoretically, from the inhibitory effects that they exerted in supporting special functions under normal condition. But this was ruled out by results of experiments where we separated mechanically the mesenchyme and epithelium to enable the epithelial cell to proceed and grow indefinitely.

DR. GIRARDI: I wanted to add a preliminary note about the complement-fixing antigen in the tumor cell and its relationship to transformation. In an attempt to try to rescue the SV40 viral genome from cells we studied human cells transformed by SV40 virus which were no longer shedding infectious virus. We went through many experiments which failed. Then we attempted to extract a nucleic acid from the transformed cell and use this nucleic acid in the sensitive lytic system to try to recover virus. Since the nucleic acid just happened to be available and since we had the parent cell which represented the original cell that was transformed by the virus, the nucleic acid from the diploid transformed cell was placed in the normal human cell culture. When the cells were analyzed they were shown to contain the neoantigen for SV40. The experiment was repeated with a new batch of SV40-transformed cells and a new batch of nucleic acid. Again, going back to the normal diploid cells, the neoantigen was made in these cells, and the cells did not go on to become "immortal." The difference of measuring units for transformation and development of antigens can be examined by this method. I think it is interesting to possibly separate these two, and with the nucleic acid one might be able to exchange this information from one cell to another.

DR. SABIN: I want to be sure that I understood Dr. Girardi very clearly, because this is quite a new finding, and it is a little bit surprising. You are saying that human cells which have been transformed by SV40 virus, and no longer produce infectious virus, did produce the complement-fixing T antigen; nucleic acid extracted from the transformed cells and added to other cells did not transform them to malignancy, nor to continued growth, but *did* result in the capacity to produce the T antigen, which could then be carried on in subsequent generations, or just one?

DR. GIRARDI: In subsequent generations.

DR. SABIN: This is very interesting, because when we attempted to carry out similar tests with the nucleic acid extracted from hamster tumors which had the SV40 virus genome and which could be reactivated by association with living cells, we were unable to produce any infectious virus by adding that nucleic acid on hamster embryonic tissue. But we have never tested for the T antigen. If this is correct, and I think this is a very important departure, Dr. Girardi may have separated here the genetic information required for the synthesis of the T-antigen from the genetic information required for producing transformation. I understood you correct, is that right?

DR. GIRARDI: Right.

DR. SABIN: Thank you very much. I think that is a very important observation, and I hope we all repeat it.

DR. GIRARDI: I hope you can, too. In fact, I am the most skeptical about this of the group that is involved. Drs. Roland, Maes and Antti Vaheri at the Wistar Institute have done the extractions from the cells I supply. If this holds up through very rigorous testing, I think it is interesting. There is one point that makes a difference: the cell that we are going back to with the nucleic acid is actually the parent of the cell that has become transformed. We were able, through liquid nitrogen storage, to store the parent cell, in this case the WI26 line. We are trying now to do the same with the WI38 which is of different sex and from a different embryo. It may not work under those conditions.

DR. HUEBNER: Some of our discussants gave papers longer than the speakers', and I figure they ought to have a chance here too, Dr. Black.

DR. BLACK: Dr. Sabin, there are now reports from two laboratories, Dr. John Enders' and ours, in which cells became infected with SV40 virus and DNA from SV40 respectively. In the first case, cell lines were established, and in the second case, a malignant potential was imparted to the cell line. In Dr. Enders' case, he infected with SV40 virus and he gets increased longevity of cells from hamsters in which the cells have become lines and there is no SV40 T-antigen. In our case, we infected the BHK 21 line with SV40 DNA, and a new malignant potential was conferred to these cell lines. There is absolutely no evidence of SV40 DNA in some of these cell lines; there is no T-antigen; there is no virion antigen; there is no transplantation rejection antigen, and there is no DNA recovered by hybridization, that is, after we make an *in vitro* RNA and take this SV40 m-RNA and try to rescue DNA from cells. We are postulating that a part of the SV40 genome is conferring a malignant potential to these cells that we cannot measure by a marker or cannot rescue, or may not be there at this time. Now, what Dr. Girardi may have gotten from his cells is that portion of the SV40 DNA which codes for the T-antigen and we think this portion resides in all transformed cells which have the T-antigen. It may have nothing to do with the malignant potential of these cells. It may just be that cistron that codes for T and this does not have to be linked with the malignant potential. There are evidences of transformed cells and tumors which have their malignant potential perpetuated but have lost their T-antigen and transplant-rejection antigen. There is a paper from Russia that is coming out on this and there is another cell line that has lost its T and is still malignant.

DR. SABIN: This is a very interesting situation. Now, what Dr. Black was saying is that with nucleic acid extracted from fully infective virus you can transform cells. Well, that is all right. No argument there, because you are reproducing the whole thing. But the other thing that he is saying, about which I think we must be extraordinarily careful, is that you can get the malignant transformation (and I use that for tumors, for cells that can be explanted, as of hamsters, and prove that they are malignant), and yet not get the synthesis of the so-called tumor antigen. I went through that with polyoma virus which had been cloned by these beautiful colonies of BHK cells by Dr. Kingsley Sanders, and it was still having all of the properties of transformed cells. I tried to demonstrate T-antigen in it, and I couldn't. And since I knew from many experiments that different strains of polyoma virus can produce different kinds of T-antigens, I tested the serum of the tumor-bearing hamster against its own tumors. Still, there was no antibody. And so, I could have reached the conclusion that Dr. Black has just made. But I was able to show that that was an erroneous conclusion because it was merely a situation where the amount of T-antigen synthesized in this tumor cell was so small that it could not be detected. But, it was enough so that when the tumor was grown in another hamster, it produced antibody which, when tested against polyoma tumor that had plenty of this antigen, could be demonstrated to have that particular antibody. Therefore, I think I would like to ask, for the record, before I write the summary: I know Dr. Black is familiar with this phenomenon which Dr. Huebner has demonstrated, for example, that there is not enough c-antigen, viral antigen, in adenovirus-12 tumors to be able to demonstrate its presence by using an anti-c serum. And the only way that you know it is there is because the tumor-bearing hamsters, after the tumor has grown for a long enough time, produces antibody for c. Now, Dr. Black, would you please tell me whether in the instances where you said there was no longer a T-antigen, whether or not you tested the serum of hamsters in which the tumors had grown a long time against other antigens, please.

DR. BLACK: Dr. Sabin, when you say no longer—what I just said about the loss of T-antigen or transplant-rejection antigen in malignant neoplasms propagated *in vitro* or in animals, that was work from Russia. The cell lines I referred to never had T-antigen demonstrated in them. They were DNA "transformed"—they had an increased malignant potential conferred on them which was genetic, by SV40 DNA. None of these cell lines had T-antigen; when the cells were transplanted into animals they had at least a 2 log more malignant potential than the parent line. The tumors contained no T-antigen; the animals developed no T-antibody. Throughout ten *in vivo* passages no tumor antigen was ever found in any of the passages; no tumor antibody was ever found in any of the animals. Also, and I think more sensitive than the T-antigen is nucleic acid hybridization. Whenever we find T-antigen, we can hybridize SV40 DNA from cells. We regard this as evidence that the T-antigen is formed from SV40 DNA and stems directly from the genetic information contained within the SV40 DNA and is not from derepressed host cell genetic information.

DR. SABIN: How long after the tumors were grown did you obtain the serum from the hamsters?

DR. BLACK: Until they were big. Dr. Sabin, we were the first to show the correlation between T-antigen, T-antibody formation, and size of tumor, so we are well aware that the T-antibody concentration in the hamster is directly proportional to the size of tumor. When one excises the tumor, T-antibody falls.

DR. RAPP: I think we must differentiate here between cells exposed to, and cells infected by, or with, viruses. Take Dr. Macpherson's very beautiful BHK21 cell system which requires 10^5 or 10^6 cells for induction of a tumor. In our laboratory, ten BHK21 cells will make that tumor, and those cells have not knowingly been exposed to oncogenic viruses, nor can we measure the presence of any of the known T-antigens. Another hamster cell in our laboratory appears to be oncogenic, again, without exposure to any virus. We cannot detect any of the known T-antigens, and the cells making the tumors do not cause development of antibodies against any of the known T-antigens. These cells apparently have immortality as measured by some 90 passages in tissue culture. So, I think that if we had treated those cells, for instance with SV40, and transformed them without subsequent synthesis of the T-antigen, we would be unable to decide whether SV40 had or had not transformed the cells. I think we have to be careful to delineate between exposure versus infection.

DR. BLACK: Now this cell system, Dr. Sabin, needs the agar technique that Dr. Macpherson has referred to. When we infect BHK cells with SV40 DNA, 0.1% of them plate in agar and 0.1% of them form T-antigen. So, all the cells that grow in agar we think have the T-antigen originally, so we have evidence of direct SV40 DNA infection. Of these 0.1%, only 15% have persistent T-antigen, and their clones have T-antigen in their tumors while the other 85% lose their T. We may conjecture that the cell became infected with DNA and could thereby grow in agar. Then the viral genome either was lost, or if not lost, was not detected by any of the markers we use. I would like to report recent evidence with the 3T3 cell line of mouse fibroblasts, the line that was originated by Todaro and Green. With high titered SV40 virus, we can transform 50% of these cells. 100% of these cells were infected with SV40 virus, and they formed SV40 T-antigen. The 50% that are transformed keep their T-antigen. We presume that the virus is integrated conferring the genetic stability of T-antigen formation to these cells. The other 50% lose their T-antigen after perhaps one to two divisions. We are in the process of determining that. This is a very interesting phenomenon, we feel, because it means the virus coded for messenger RNA which was translated into T-antigen or protein, and then it was lost, that is, not integrated. So here is some evidence where SV40 entered the cell, coded for T-antigen and then exited. Either the DNA was hydrolyzed or something else, and transformation did not occur. But it is evident that the DNA entered and did something, but did not transform. In the other case, we think the DNA got in and conferred a growth stimulus to the cell which resulted in an agar colony but soon after was lost by some, as yet, unexplained means.

DR. HABEL: I think that the question that Dr. Sabin brought up is a very important one: that is, what does a negative test mean? If a cell is negative for T-antigen, it just means that it has less than the minimum amount that is necessary to be positive. It does not mean that it is absent. I think the same thing is true when we were talking about oncogenesis: which adenoviruses are oncogenic, which cells produce tumors? All you can say is that under the conditions in which you are testing for its ability to produce a tumor and in a particular experimental setup it is negative. And I think that one of the possible implications of this transplantation antigen which we have talked about today is the fact that the very cell that does not produce a tumor when you transplant it into an isologous host may be the one that is most transformed from an antigenic standpoint. Maybe the reason it does not produce tumors is because it is rejected immunologically much more efficiently than the one that does produce tumors. Therefore, unless you use special methods like whole-body x-ray or thymectomized animals, and go to some extreme to rule out the immunological aspect of this, you cannot say that one thing is oncogenic, or not oncogenic, under a given set of circumstances.

DR. LASFARGUES: In our system, we have embryo mouse cells that were transformed but were not producing tumors when implanted into isologous mice for eight months. They were therefore transformed from the fifth month on and you could say that they were more malignant than the ones we had during the eleventh month.

DR. HUEBNER: Thank you.

Summary of Symposium on Malignant Transformation by Viruses[1]

The Children's Hospital Research Foundation,
University of Cincinnati College of Medicine,
Cincinnati, Ohio

The purpose of a summary of a symposium such as we have just attended is obviously not merely to give a series of ultra-brief abstracts of everything that has been presented, but rather to give one man's perspective of the bearing that the data to which we were exposed during the last two days have on some of the fundamental issues in the malignant transformation of cells by viruses. Another question that I have had to consider was for whom is such a summary intended. Is it for my many colleagues, who meet each other rather frequently at such symposia critically to evaluate new advances and to be stimulated into new lines of experimentation, or is it for the benefit of people who are engaged in work on malignant transformation of cells by procedures and substances that do not, at least knowingly, involve viruses, or is it for the undergraduates at this symposium, who may be enticed to enter the field of cancer research, because the program indicates that the symposium is sponsored by an Undergraduate Cancer Training Grant from the National Cancer Institute? I have decided to attempt a summary that may be both comprehensible and meaningful to all of these people.

Now, why are we interested in malignant transformation by viruses, or rather, from what point of view are we interested in this problem? Well, first of all, as biologists, or "pure" medical scientists, we are curious to know what happens when a normal cell is transformed to malignancy by a virus. A man can be a perfectly good scientist and yet hate human beings; he may not care at all about the significance his findings might have for human cancer and still make excellent contributions. Some of us, however, in addition to being endowed with this basic curiosity, are also "humanitarian" medical scientists, and every bit of knowledge that emerges from basic studies we are anxious to apply to human problems. And so, (I'll class myself as a humanitarian medical scientist) whenever any work is presented dealing with the basic biologic differences between normal cells and malignant cells I have my brain cells alerted to "what possible use can this bit of knowledge be in trying to answer the question which still remains unanswered: whether any human cancer is caused by a virus."

The organizing committe decided that this symposium should deal with three aspects of malignant transformation by viruses: 1) *characteristics of malignant transformation,* which I interpret to mean how, or in what way, does the malignant transformed cell differ from the normal one? 2) *genetics and immunology of malignant transformation;* in other words, particularly in the light of the papers that were presented at this symposium, what does the viral genome do and what happens to it in a cell that it has transformed, and what does it do to the genetic material of the cell? And then, what are the changes that lend themselves to study by immunologic methods?, and 3) *the significance of malignant transformation in relation to human neoplasms,* which is the thing that obviously has the humanitarian overtones that I just posed: "What can you do with the currently available knowledge to elucidate the problem of the role of viruses in human cancer?"

Characteristics of Malignant Transformation

Biological manifestations of cells transformed to malignancy by viruses. Dr. Macpherson in the very first paper made reference to contact inhibition of cells that are transformed to malignancy, and he illustrated the very interesting procedures that they have been using in Glasgow: the differences of colonial morphology of individual cells that are grown in Petri dishes, and especially the interesting use that they have been able to make of the procedure developed by Kingsley Sanders in which cells are put into soft agar, where non-transformed cells will not form colonies, while the cells transformed to malignancy *do* form colonies. It has certainly been a very useful tool in the hands of Dr. Macpherson, Dr. Stoker, and the fine group working in Glasgow, but the question that arises is its applicability to other transformed cells, and even to the cells which they use themselves, because at least some viral oncologists working in this field are troubled by some of the properties of the BHK21 cell line. Dr. Macpherson told us that the starting cells which do not exhibit the *in vitro* growth characteristics of malignant cells mentioned above, nevertheless, show malignant properties *in vivo* when a million cells are inoculated in hamsters. The only difference then, between the BHK cells used for transformation and those transformed by viruses, as regards their oncogenic activity in hamsters *in vivo,* would appear to be a quantitative one. Certainly, you cannot call the BHK21 line of cells *normal* cells; you are not in this case transforming a normal cell to malignancy; you are transforming a cell that already has a certain oncogenic potential to greater oncogenic potential. And this has been shown to happen spontaneously without any known viruses, but unquestionably more quickly and regularly and beautifully, as Dr. Macpherson has indicated, when certain viruses are added to these cells.

Dr. Temin's report mentioned a number of interesting new properties that chicken cells acquire when they are converted *in vitro* by infection with Rous sarcoma virus: the greater glycolysis, greater synthesis of an enzyme, the hyaluronic acid synthetase, the greater production of acid mucopolysaccharides; and he mentioned still another very interesting phenomenon which particularly affected my "humanitarian" antennae. He referred to the phenomenon of the requirement of a

serum factor for cell division; and he showed that the chicken cells that are
transformed by Rous sarcoma virus, unlike transformed mammalian cells, have a
finite life; they will *not* keep on multiplying as long as you give them proper
medium; the main difference between the transformed cells and untransformed cells
is that when you cut down the amount of the "serum factor" the normal cell
will stop multiplying whereas the converted cells can continue to multiply for a
much longer time. Now, why did my "antennae" react to this? Because in work
with material from human tumors, many of us have tried to see whether by addition
of certain tumor extracts you can make a normal cell, which has a finite period of
growth, to grow progressively even though without obvious morphologic transformation; and what most of us have usually done is to provide the best medium
for it. We then find one of two things: either that the control normal cells as
well as the "treated" cells continue to grow perfectly well, without any early
indications of a difference in finite life, or that they both stop growing. What I'm
wondering now is whether those of us who are engaged in this kind of work may
not have in Dr. Temin's observation, still another possible approach: when you
have treated human cells with material potentially or hypothetically containing a
human oncogenic virus (viruses from leukemia or lymphoma where you may
expect to find infectious virus) and you are looking for changes which are not
expressed morphologically, perhaps the reduction in the amount of serum in the
medium could serve to indicate possible transformation if it should stop the
multiplication of the control cells but not of the "treated" cells. If it doesn't work
I'll be the last one to be surprised because the history of those of us who are
engaged in trying to apply various procedures that have been observed in model
systems to the human problem has been that we have tried it, but it didn't work.
But one of these days, something may work.

Before going on to another aspect of biological manifestations of malignant
cells, I wish to say that one of our very great needs is some biochemical or other
sign of malignant transformation of cells by those oncogenic RNA viruses which
at the present time can be detected *only* by the production of tumors in the animal.
For example, in the case of murine leukemia, if we didn't have newborn animals
of the proper breed to inoculate, we would still not be able to demonstrate that it is
caused by a virus. Therefore, some *in vitro* system, such as Dr. Temin has described,
would be extraordinarily helpful.

Doctors Nichols, Moorhead and Yerganian presented data bearing on the question
of chromosomal aberrations as indicators of malignant transformation. This is a
field in which I have very little competence (I think it would be more correct
to say I have no competence), and therefore, I was in the position of other members
of the audience in relation to other spheres of competence. I found that I had
great difficulty deciding what the data presented by these people really meant, so
I got them aside last night for a little additional personal education. I am greatly
indebted to Dr. Nichols and Dr. Yerganian (Dr. Moorhead was already gone) for
giving me the following estimate of the situation. Dr. Nichols pointed out that
when the Schmidt-Ruppin strain of Rous sarcoma virus converts mammalian cells,
one can find chromosomal breaks and chromosomal fusion. But he stated that since

similar effects have been produced by non-oncogenic viruses, as Dr. Rapp pointed out, and since similar chromosomal aberrations can be produced by inhibitors of DNA synthesis without inducing malignancy, the practical usefulness of such chromosomal aberrations is, therefore, uncertain. This effect of inhibitors of DNA synthesis on chromosomes was of special interest to me because it has now been shown by several workers that DNA viruses like polyoma and SV40, which can establish only a portion of their genomes in cells that they transform to malignancy, stimulate synthesis of cellular DNA in cells in which they cannot multiply but can transform. Dr. Moorhead described a variety of chromosomal aberrations in human cells transformed by SV40 virus and Dr. Yerganian stressed how the presence or absence of chromosomal aberrations, as well as the kind of aberration, can be influenced by the species of animal cells. He called attention to his very interesting studies on the Chinese hamsters and the Armenian hamsters, which are rather new to laboratory experimentation. When I asked Dr. Nichols and Dr. Yerganian whether, in their judgment, at the present time, chromosomal aberrations *by themselves* can serve as indicators of malignant transformation in the absence of other signs, they said "no."

Factors influencing malignant transformation by viruses. A number of the participants made important observations on this aspect of the problem. The report of Dr. Dawe and his associates was of particular interest to me both because of the methodology and the conclusion that "epigenetic" factors were involved in oncogenesis by polyoma virus. They worked with organ cultures of fully formed or embryonic rudiments of the submandibular salivary gland of the mouse, which were infected with polyoma virus *in vitro,* and the neoplastic character of the transformation was established by transplantation to syngenic newborn recipients. They established that adult, newborn and embryonic rudiments (14-day embryos) were equally responsive to the oncogenic action of the virus when tested by this technique. The point of special interest was that when the epithelium and the mesenchyme of the embryonic submandibular gland were separated by dissection with the aid of a little trypsin, and the two portions were infected separately, no transformation occurred and no tumors appeared in the transplants of the separate components. That the manipulation itself was not responsible for the negative results was proved by the fact that when the infected, separated components were recombined neoplastic transformation occurred and typical salivary gland tumors appeared on transplantation to the syngenic newborn mice. The specificity of the submandibular gland epithelio-mesenchymal interaction was established when it was shown that substitution of either component by either the mesenchyme or epithelium from tooth-germ, failed to yield neoplastic conversion. Thus it was established that for neoplastic conversion of the 14-day mouse embryonic submandibular gland epithelium, a specific mesenchymal "helper cell" was required—although just how the helper cell achieved this effect was not elucidated.

Dr. Kirsten's data on the effect of polyoma virus on organ cultures of embryonic mesenchyme and epithelium from rat kidneys, while being different from those of Dr. Dawe in that both the mesenchymal and the epithelial rudiments infected separately exhibited neoplastic transformation—each resulting in a different type of tumor—nevertheless demonstrated an interesting effect on the nature of the resulting neoplastic cells when the mesenchymal and epithelial rudiments were growing together but separated by a membrane.

It occurred to me that the interesting data presented earlier in the symposium by Dr. Howard Green might have a bearing on the interpretation of the data presented by Doctors Dawe and Kirsten. Todaro and Green had just reported (*Proc. Nat. Acad. Sci., U.S.A.,* 55:302, February, 1966) that for the 3T3 mouse cell line to be transformed by SV40 virus, at least one cell generation was required for establishment of the transformed state as a heritable character and that several more generations were required for the "expression of the transformed state." They stated that "cells which are not permitted to grow through a cell division cycle subsequent to infection are altered neither in their genotype nor in their phenotype."

I have had the benefit of a further discussion of this question last night with Doctors Dawe, Kirsten and Green. In answer to the question whether the rudimentary mesenchymal cells and epithelial cells from the 14-day mouse embryo submandibular gland multiplied in organ cultures, Dr. Dawe said that they did not when they were planted separately but that multiplication and differentiation did occur when they were planted together. It is possible, therefore, that the main influence of the mesenchyme in Dr. Dawe's experimental system was to induce multiplication of the epithelial cells and thereby, in the light of the work of Todaro and Green, make it possible for the polyoma virus to induce transformation. While it is a possible explanation, it may not be the real explanation of the interesting phenomenon reported by Dr. Dawe. It is noteworthy, however, that Dr. Kirsten stated that his rat kidney embryonic mesenchymal cells and epithelial cells did multiply when planted separately in organ cultures—an observation that appears to fit the requirement of cell multiplication as a prerequisite for transformation.

Drs. Defendi and Girardi reported interesting observations on the role of age, as measured by the "finite life" of a cell line, on morphologic transformation of cells by SV40 virus. Convincing data were presented that the older cells (judged by the above criterion) are much more rapidly transformed than the younger ones. One is tempted to speculate on the bearing that these observations may have on the fact that, with the exception of certain juvenile malignancies, cancer in man is a disease of advanced years. Moreover, we still have no explanation why so many months often elapse between the inoculation of an oncogenic virus in newborn animals and the appearance of a tumor. We know that the viral genome gets into the cells quickly but the events that occur during the long interval before a tumor appears in the inoculated animal are still shrouded in mystery.

The data reported at this symposium by Doctors Lasfargues and Moore are of special interest because they deal with the mouse mammary tumor virus, which thus far still can be recognized only by transmission to appropriate newborn mice. The development of a technique for its recognition in tissue culture *in vitro* would be a great achievement that would provide new approaches for the search of a possible viral etiology of human breast cancer. As I understand it, mammary cells from virgin, virus-free Ax mice were grown in tissue culture—not in organ culture. Such cultures, Dr. Lasfargues told me, at first consist of a mixture of epithelial cells and fibroblasts, and after a while only undifferentiated fibroblast-like cells remain. An untreated culture and a culture that was initially exposed to a large concentration of mouse mammary tumor virus from strain A mice were kept in continuous passage and subcultured when necessary. The untreated culture had a finite life of about 4 months, while the virus-treated culture is now in its third year of propagation and still growing. We were told that transplantation of these cells subcutaneously and into the mammary gland of the isologous Ax mice produced no tumors—so there is no evidence for malignant transformation. But is it possible that the longevity itself is due to the exposure to the mammary tumor virus? Since mouse fibroblast-like cells can occasionally be established as "permanent" lines without the "known" help of a virus, it is obviously important to know how often one can repeat the observation that control cells will die in 4 months or less while mammary tumor virus treated aliquots of these cells will exhibit the reported longevity.

In the second experiment reported by Lasfargues and Moore, C57 mouse embryo cells, exposed to milk from the RIII strain of mice, have now grown for 14 months. In this case, however, transplantation of cells after 8 to 9 months of cultivation *in vitro*, produced tumors in mice—but these were sarcomas, not mammary carcinomas. Again we must ask whether the mammary tumor virus in the milk had anything to do with the neoplastic character of these cells? The milk could have contained another virus or the cells may have undergone spontaneous transformation to malignancy. What do you do in a situation like this? Just let it go because of the various other possibilities, or set up more extensively controlled studies that will permit a definitive answer? I favor the second alternative because the issue is so very important, and because there is always the possibility that the differentiated mammary gland cells may not be the only target cells for this virus. At the same time that such work is being followed up, it would seem to me important to work more intensively with organ cultures of mammary glands, that are stimulated to multiply by hormonal or other means, to see whether under such conditions the mammary tumor virus may exert an oncogenic or other distinctive effect *in vitro*. I am told that this is not an easy field in which to work, that results come slowly, and that efforts are more likely to meet with failure than success. It seems to me, therefore, that it is all the more deserving of the attention of more mature investigators who can more readily afford to gamble a number of years on potentially unsuccessful explorations. I am somewhat depressed by the impression that the more money becomes available for research, the more many people are inclined to do the easy, often inconsequential but publishable, work. Perhaps only concentrated effort on the more difficult approaches will ultimately yield the answers we are looking for.

Genetics and Immunology of Malignant Transformation

So many of the participants in this symposium—Macpherson, Temin, Rapp, Black, Habel, Huebner, Girardi and others—have presented data bearing on this topic that I will not find it possible to evaluate their contributions separately.

Instead I should like to examine our current knowledge in this field under separate headings.

What happens to the virus in cells transformed to malignancy? The answer to this question is different for the RNA viruses responsible for the *naturally occurring* malignancies and for the DNA viruses, with which virologists have artificially produced cancers. In the cells of their natural hosts, the RNA oncogenic viruses, i.e., those of Rous sarcoma and leukosis of chickens and of mouse leukemia and mammary carcinoma, can continue to multiply with the production of fully infectious virus. This fact is responsible for the easy demonstration by routine virologic procedures of transmission to appropriately susceptible hosts that these naturally occurring malignancies are caused by specific viruses. It is noteworthy that these oncogenic RNA viruses have not been found to have any cytocidal effects in any cells thus far tested; they either multiply without any recognizable morphologic effects or they produce malignant transformation. *In vitro* transformation of susceptible chicken target cells has thus far been obtained only with the Rous sarcoma and avian myeloblastosis viruses, and not with the other leukosis viruses. The special case of Rous sarcoma virus, which needs the help of certain avian leukosis viruses for replication of infectious particles, and the experimental production of malignant chicken cells which contain only the genome and other noninfectious products of the Rous sarcoma virus, have been well documented. The interesting demonstration that Rous sarcoma virus strains possessing certain "outer coats" (presumably with receptors for the cells in question) can convert certain mammalian cells (e.g., hamster, rat, monkcy, human) to malignancy *in vivo* or *in vitro* or both, and that under such conditions the viral genome continues to be transmitted from generation to generation without the production of infectious virus until appropriate contact is made with chicken cells and certain avian leukosis viruses, is still another confirmation that the viral genome of the naturally occurring RNA viruses persists in the malignant cells that they produce.

The situation with the experimentally oncogenic DNA viruses is in some respects different from, and in others similar to, that obtaining in mammalian cells *experimentally* converted to malignancy by Rous sarcoma virus. To begin with, it must be said that all the DNA viruses with which experimental malignancies have been produced, are cytocidal for some cells either in the host in which malignant tumors can develop, e.g., polyoma virus in mice, or for cells from other hosts, e.g., SV40 in African monkey kidney tissue cultures or the oncogenic, human adenoviruses in various human cell lines. Yet it can probably be said that, as a rule, no infectious virus is produced in the cells that are transformed to malignancy by these DNA viruses, except under the special conditions to be mentioned later. It can also be said that there is no need for these DNA viruses to multiply to the point of production of infectious viruses in order to transform cells to malignancy. This has been amply demonstrated for the SV40 virus and some adenoviruses both in the living newborn hamsters and in hamster embryonic tissue cultures. Since it is now generally accepted that at least a part of the genome of the oncogenic DNA viruses persists in the malignant cells and continues to be transmitted frcm generation to generation with each cell division, I should like to examine the evidence on which this conclusion is based.

Evidence for continuous transmission of viral genome in cells made malignant by DNA viruses. While the specific virus-induced transplantation resistance phenomenon in adult animals, described in detail at this symposium by Habel, is probably due to the production of new antigens by continuously transmitted viral genome in the malignant cells, this phenomenon by itself does not, in my judgment, provide evidence for persistence of viral genome. Why? Because there is as yet no evidence that the antigens involved in this phenomenon are a regular product of the viral genome in all infected cells and it is conceivable that the transforming virus could have produced a specific heritable change in the original cell without persistence of the viral genome in the progeny. What then can be considered as the first evidence that viral genome persists in the transformed cells? In my opinion it is the phenomenon of *discontinuous* release of minute amounts of infectious virus by continuously transplanted cells of certain tumor lines. While there was a suggestion that this may occur in certain lines of polyoma tumors, it was difficult to prove it because the polyoma virus can multiply in so many different types of cells and because clones of cells could be obtained which failed to exhibit this phenomenon. Proof, however, could be obtained with certain lines of SV40 tumor cells because the SV40 virus does not produce infectious virus in hamster cells. The critical experiment involved the demonstration that tumors resulting from only one or two cells (representing at least a 100 millionfold dilution of the original, washed, trypsinized hamster tumor cells) yielded traces of infectious virus with the same frequency as tumors (Eddy strain) resulting from transplantation of 10^6 to 10^7 cells (see Sabin, A. B., and M. A. Koch, *Proc. Nat. Acad. Sci., U.S.A.,* 49:304, 1963; *ibid.,* 50:407, 1963). These results permitted the interpretation that all, or almost all, cells of this strain of SV40 tumor contained enough *repressed* viral genes to permit an occasional cell to yield infectious virus under special conditions of *derepression*—a situation similar to that obtaining in *nondefective* lysogenic bacteria. Using such SV40 tumor cells it was, moreover, possible to increase the number of cells yielding infectious virus by a variety of procedures which included prolonged serial cultivation *in vitro,* radiation with X-rays, exposure to proflavine, hydrogen peroxide, mitomycin C, and especially by the phenomenon of "induction" by contact of live tumor cells with normal African monkey kidney cells under special conditions of cultivation. In a typical experiment involving "induction" by contact with normal susceptible cells, many millions of washed, trypsinized, frozen, and thawed tumor cells fail to yield infectious virus, whereas an aliquot of the living cells from the same tumor yields *minute* amounts of infectious virus when 10^6 and occasionally smaller numbers of cells are used. Moreover, starting with tumor cells, of which a minimum of 10^6 cells was required for induction of development of only one to ten infectious doses of virus, continued serial cultivation of the tumor cells altered the inducibility to a point where only 10^2 to 10^3 cells were sufficient to yield infectious virus. The special conditions required for "induction" by contact with normal cells are of interest because they suggest that some kind of cytoplasmic association may be required. Thus, in our studies (Sabin, A. B., and

M. A. Koch, unpublished data) the optimal conditions of "induction" by contact involved the use of well spread-out multiplying normal cells in tightly stoppered tubes or bottles in which the tumor cells could spread out adjacent to the normal cells. When we used tightly packed cell sheets under fluid medium in bottles, or islands of normal cells under agar, there was, as a rule, no "induction." As a further analogy to nondefective, lysogenic bacteria, it proved impossible to extract *infectious* viral DNA from many millions of such SV40 tumor cells.

Although with the Eddy strain of SV40 tumor it was found that even after 68 serial transplantations in hamsters over a period of several years the cells retained repressed viral genome capable of induction by appropriate contact with susceptible, monkey kidney cells (L. Geder and A. B. Sabin, unpublished data), it has proved impossible to demonstrate such repressed viral genetic material in several other lines of SV40 tumors, polyoma tumors and adenovirus hamster tumors. The question arose, therefore, whether all of the viral genes had disappeared from these tumors or whether they merely became defective, comparable to defective lysogenic bacteria which retain some viral genes but cannot be induced to form infectious virus.

The first indication that nonvirus-yielding tumor cells still possessed viral genetic information, responsible for the synthesis of some *virion* components but not of infectious virus, came from the studies of Huebner and his associates on the hamster tumors produced by adenovirus type 12. These investigators showed that hamsters bearing these *transplanted* tumors for long periods of time developed *type-specific* neutralizing antibodies for adenovirus type 12 (and this to me was the crucial evidence for persistence of viral genome in the nonvirus-yielding tumor cells) and complement-fixing (CF) antibodies for the purified, type-specific "C" antigen contained in the viral particle. It is noteworthy that the amount of "C" antigen present in the tumor cells was too small to be detected by a CF test with specific "anti-C" rabbit serum. They also demonstrated that sera obtained earlier after transplantation of these tumors in hamsters, which were devoid of neutralizing antibodies and of "anti-C" CF antibodies, nevertheless, exhibited specific CF activity for extracts from the tumor cells. Since similar specific nonvirion antigens were demonstrated in nonvirus-yielding SV40 and polyoma tumors, and since no neutralizing or other *antivirion* antibodies could be demonstrated in the sera of hamsters bearing these tumors, as well as adenovirus type 3 and 7 tumors, for long periods of time, the question arose whether these specific, nonvirion tumor CF antigens (subsequently also demonstrated by the fluorescent antibody technique to be present in the nucleus of every tumor cell) were also products of persisting viral genome. The fact that the Fortner Sarcoma No. 3 (F Sa 3) hamster tumor, which originally appeared at the site of a sodium cholate injection and in which we have been unable to demonstrate either a passenger virus or mycoplasma, also turned out to have a specific antigen, which stimulated an antibody in tumor-bearing hamsters that did not react with the virus-induced hamster tumors (Sabin, A. B., and L. Geder, unpublished studies), suggested that the mere presence of a specific tumor antigen could not be taken as evidence of persisting viral genome. Nevertheless, it was possible to prove that the specific tumor antigens were identical with nonvirion antigens produced early during the course of cytocidal infections of cells with these viruses, and, therefore, also a product of specific viral genes in the tumor cells. In the case of SV40, it was possible to show that all of the specific anti-tumor CF antibody in the serum of tumor-bearing hamsters could be completely absorbed specifically by the products of SV40 virus infection in normal cells.

The question also arose as to whether the specific tumor antigens coded by viral genome are the only "new" antigens in the tumor cells, because we have found that a varying proportion of hamsters bearing transplanted, virus-induced tumors for 5 weeks or longer develop CF antibodies that react not only with tumors induced by other viruses but also with the F Sa 3 hamster tumor of no known viral etiology. It is possible that this may represent a response to certain isoantigens that may be present in *increased* concentration in the tumor cells, because some but not all hamsters yielding such cross reacting sera also *develop* very low titers of CF antibody for normal hamster kidney or embryonic tissue. However, since some high-titered, cross reacting sera react only with tumor cells, the possibility of some common tumor cell antigen cannot as yet be eliminated.

In conclusion, it can be said that the available data indicate that the amount of viral genome, as reflected by its phenotypic expression, that persists in transformed cells is variable: all cells possess the portion that gives rise to at least some of the nonvirion "early" antigens; some (i.e., adenovirus type 12 tumors) in addition contain the portion required for the synthesis of at least one virion antigen, and others (e.g., some lines of SV40 tumors) appear to have a complete "provirus" in a "repressed" state which under certain conditions of spontaneous or induced de-repression can produce fully infectious virus in a varying proportion of the cells.

Some properties of specific tumor antigens and appearance of antibodies against them in tumor-bearing animals and after ordinary infection. The immunology of malignant transformation, especially by the DNA viruses, offers certain approaches to determining whether any *naturally occurring* tumors of human beings and lower animals are caused by any of the known viruses of this group. It is important, therefore, to know in considerable detail the variations in the properties and quantitative aspects of the specific tumor antigens in the available experimental model systems. Because of limited time Dr. Habel could make only certain general statements that are not entirely true. For example, it is often stated that the specific tumor antigens are "soluble" when it is meant that they are not sedimented by centrifugal speeds that remove the virus particles. Centrifugation for several hours at more than $100,000 \times g$ can remove as much as 75 per cent or more of the CF activity from the supernatant fluids, and different results can be obtained when tumor extracts or infected normal cells are used. It is often said that the CF activity of the specific tumor antigens is inactivated by $56°$ C for 30 minutes while that of the viral particles (the virions) is not. While this is true for all SV40 nonvirion, tumor antigens tested it is not true for at least one line of polyoma or for some of the antigens in adenovirus 12 tumors when certain sera are used. The effect of ether is also different: the CF activity in adenovirus tumors has been reported to be unaffected, while 90 per cent of the activity is lost in the SV40 and HE-11 strain of polyoma tumors, and 75 per cent in the H-1 strain of polyoma tumor. We must also realize that we are not dealing with single antigens but rather with multiple antigens for each tumor line and that antibodies for all of the antigens may not appear in all tumor-bearing hamsters or may appear at different times. Studies in my laboratory on three different lines of polyoma hamster tumors showed that: 1) many hamsters bearing the H-1 line produced CF antibodies that reacted only with H-1 but not with the HE-11 tumor, 2) sera from HE-11 tumor-bearing hamsters invariably reacted with both HE-11 and H-1 tumors, and 3) a cloned Stoker-Sanders tumor line had so little of the specific tumor antigens that no reaction was demonstrable with homologous, HE-11 or H-1 sera, although the sera from the hamsters bearing transplants of this tumor line for many weeks did react specifically with the HE-11 antigen. The latter finding is of great importance because it shows that the amount of specific tumor antigen in various lines of tumor cells can be highly variable and occasionally so small that the only way to test for it, is to look for antibody in the sera of many hamsters that have carried the tumor for many weeks.

Another generalization that is unfortunately often not true is that anti-tumor antibodies regularly appear after the tumor has grown to a "large" size. For some unexplained reason there is a great deal of variation in the appearance of anti-tumor antibodies in hamsters bearing different SV40 tumor lines, despite the fact that all of them grow rapidly to large size and the tumor cells contain high concentrations of specific tumor antigen. Almost all hamsters bearing the Eddy SV40 tumor develop CF antibodies within 21 days after transplantation, while about 80 per cent of hamsters bearing very large tumors of the Cincinnati and Melnick strains die in 4 to 6 weeks without having any demonstrable antibody. Hamsters bearing certain polyoma tumors are notoriously poor antibody producers, and Habel has obtained potent sera only by excising the original tumors and hyperimmunizing the hamsters with the tumor cells.

These observations have a special significance for both experimental work and for using the immunologic approach for the study of the possible role of certain viruses in human tumors. In the first instance it points to the caution that must be employed in drawing conclusions that specific tumor antigen can disappear from a transformed cell while its malignancy remains—a point that was discussed in connection with the data presented by Dr. Macpherson at this symposium. In connection with studies on human tumors, it is obvious that failure to demonstrate antibody in a patient's serum against his own tumor cells (either trypsinized or grown in tissue culture to get away from antibody that may be present in the tumor), does not necessarily indicate either absence of antigen or of antibody.

Another generalization that is not entirely true, and can be seriously misleading, is that anti-tumor antibodies appear only in tumor-bearing animals. It is of course true that adult hamsters inoculated with very large doses of live SV40 virus—enough to give rise to transplantation resistance—do not develop anti-tumor CF antibodies (remember the virus does not replicate with the production of infectious virus in hamsters). It is also true that inoculation of live SV40 virus in green African monkeys (in whom it does multiply) can yield sera *taken at certain times* that have CF antibodies for the virion antigens but not for the nonvirion tumor antigens as it is present either in normal infected cells or in tumor cells. Yet studies in my laboratory have shown that when green African monkeys are bled *at various intervals* after infection with SV40 virus, CF antibodies for the nonvirion antigens can appear in high titer and persist for several months (Geder, L., and A. B. Sabin, unpublished data). Dr. Habel mentioned yesterday that adult mice inoculated with large doses of polyoma virus also exhibit a transitory development of antibody for the polyoma nonvirion antigens. Since the tumor-specific, virus-induced nonvirion antigens are transitorily present in infected normal cells one cannot be surprised at the appearance of antibody in some infected animals. However, the key word appears to be *transitory*, because Huebner has shown that even in tumor-bearing animals the anti-tumor CF antibody disappears within two weeks after excision of the tumor. (For a fuller discussion and bibliography of the genetic and immunologic aspects of malignant transformation by DNA viruses, see: Sabin, A. B.: Genetic Phenomena in Experimental Viral Carcinogenesis, *Amer. J. Dis. Child.*, 111: 1–10, 1966.)

Relationship between persisting viral genome responsible for synthesis of specific antigens and malignant state of the cells. The data presented by Dr. Macpherson in the first paper of this symposium indicated that from a clonal population of BHK 21/13 cells converted to "greater malignancy" by an RNA virus (Schmidt-Ruppin strain of Rous sarcoma) it was possible by further cloning to separate cell populations that failed to produce the virus-induced antigens, had no greater malignant potential than the original untransformed BHK21 cells; i.e., about 10^6 cells were required to produce tumors in hamsters, and no longer exhibited the loss of contact inhibition and the capacity to produce colonies in soft agar. He called these cells "revertants" and suggested that this phenomenon may indicate that persistence of the viral genome, at least that portion of it responsible for the indicated properties, is required for the neoplastic properties of the cell. Dr. Macpherson told me that he did not test the sera of hamsters with tumors produced by 10^6 cells of these "revertants," and until this is done one cannot be certain that virus-induced antigens are no longer present in these cells. That some viral genome may still be present in these "reverted" cells is suggested by his observation that, unlike the original BHK21 cells, they are resistant to further transformation by Rous sarcoma virus. It is also noteworthy that he was unable to obtain "revertants" from BHK21 cells transformed by polyoma virus. These experiments, therefore, do not provide an answer to the question that was posed, except that further work may indeed show a difference between the role of persisting viral genome in cells transformed by RNA and DNA viruses.

The data presented by Doctors Rapp and Black on the oncogenicity of so-called hybrid populations of adenoviruses containing particles of nononcogenic type 2 adenovirus and particles containing noninfectious portions of SV40 genome and adenovirus type 7 genome "wrapped" in adenovirus type 2 coats, indicate that: 1) it is not necessary to have the complete SV40 or adenovirus 7 genome to produce malignant transformation, and 2) the incomplete genomes of these oncogenic viruses carried into the cells by the outer "coat" of the nononcogenic type 2 adenovirus effected both the synthesis of specific tumor antigens as well as malignant transformation. Dr. Rapp told me that thus far it has not been possible to convert the nononcogenic type 2 adenovirus to oncogenicity by growing it in the presence of SV40 virus, but this does not alter the important demonstration that particles, containing the incomplete genomes of both the SV40 and adenovirus 7 viruses, are oncogenic.

Dr. Defendi's report, during the discussion of Dr. Habel's paper, that when SV40 virus is exposed to X-rays its capacity to produce infectious virus disappears more rapidly than the capacity to produce specific tumor antigen and cellular transformation, provides additional evidence that the complete viral genome is not needed for the initial transformation. The most intriguing data, having a bearing on this problem, were presented during yesterday's discussion period by Dr. Girardi. He reported that preliminary (as yet unpublished) experiments showed that nucleic acid, extracted from nonvirus-yielding human cells transformed by SV40 virus, when added to untransformed cells of the same line transmitted to them the capacity to produce the nonvirion tumor antigen *without* morphologic transformation or altera-tion in growth potential. If these preliminary results can be confirmed, particularly in hamster cells which can then be tested for malignancy *in vivo,* important information would become available on the possible distinctness of the portion of viral nucleic acid that is required for malignant transformation and that needed for the synthesis of specific tumor antigens.

Dr. Girardi's experiment was of particular interest to me, because for 2 years (1962 and 1963) Dr. M. A. Koch and I attempted without success to transform normal hamster embryonic cells to malignancy with nucleic acid extracted from a highly oncogenic, nonvirus-yielding polyoma hamster tumor (H-1). We employed the various tricks that have been learned in bacterial transformation by nucleic acid, and had a system by which we could detect a single malignant cell from the original tumor when it was mixed with millions of normal hamster embryonic cells. However, we did not at that time test for the possible transfer of the capacity to synthesize tumor antigen. In the light of our present knowledge about the relatively small amounts of tumor antigen in most polyoma tumors, future experiments would be more meaningful with SV40 hamster tumors that produce very large quantities of tumor antigen.

Significance of Viral Malignant Transformation in Relation to Human Neoplasm

Except for Dr. Melnick's very brief introductory remarks as chairman, almozt nothing was said at this symposium about the search for a viral etiology in human malignancies. The failure, thus far, to demonstrate a viral etiology for any human malignancy is not due to lack of effort but rather to lack of techniques by which the etiologic role of either RNA or DNA viruses could be demonstrated by methods that do not involve transmission to newborn children. Cultures of human fetal tissues offer us the only other alternative. Attempts to obtain transformation of such cultures by materials from human leukemias and lymphomas—in a manner com-parable to the *in vitro* transformation obtainable with the avian myeloblastosis and

Rous sarcoma viruses—have thus far yielded negative results, but much more work remains to be done. Attempts to obtain induction of synthesis of infectious virus by appropriate association of human malignant cells with various normal cells— along the lines of the positive results obtained with some lines of hamster SV40 tumors—have been made without success in my laboratory. Studies on the possible presence of specific CF or fluorescent antibodies or both in the serum of patients against their own tumor cells were carried out in my laboratory and in Dr. Melnick's laboratory, again without any suggestion of positive results. In the light of my earlier discussion, it can be seen why these negative results are entirely inconclusive. A cooperative study has recently been organized with the help of the National Cancer Institute to test by another approach the possible role of the DNA viruses of the human heritage and of vaccinia virus in various human malignancies—utilizing all the knowledge that has been accumulated thus far from the experimental model systems. So much work is involved in this that at least two years will probably elapse before any results are available.

In the meantime it is hoped that as more is learned about the naturally occurring RNA oncogenic viruses of mice and chickens, and about the experimental malignancies produced by the cytocidal DNA viruses, new approaches to the study of human malignancies will become possible. I sometimes ask myself how long should one continue with the search for a possible viral etiology in human malignancy in the face of repeated frustrations? The only answer that I can come up with is that as long as there are reasonable questions to ask and reasonable techniques with which to attempt to answer them, so long it is necessary to persist.